Hidden Markov Models in Finance and other Applications

This book provides fundamental concepts and algorithms of the Hidden Markov Model (HMM) and its applications in finance, such as stock price predictions, and other areas such as speech recognition. Their wide range uses make HMMs very attractive to researchers in both academia and industry. Only a basic knowledge of probability, statistics, and programming is necessary, and readers will learn the concepts and algorithms of the HMM through definitions, real-life examples, and R code.

Key Features:

- A comprehensive introduction to the concepts and algorithms of Hidden Markov Models (HMMs)
- Real-world examples that can be worked through using a calculator or R
- Applications across disciplines, including finance, bioinformatics, and speech recognition
- Fully annotated R code for hands-on learning and practical implementation

Nguyet Nguyen is an Associate Professor of Mathematics at Youngstown State University. She earned her M.S. and Ph.D. in Financial Mathematics from Florida State University and specializes in predictive modeling, data analysis, and quantitative finance. With over a decade of research on Hidden Markov Models, she has published widely on their applications in stock forecasting, economic modeling, and portfolio management, and collaborates actively with industry partners.

Hidden Markov Models in Finance and other Applications

Nguyet Nguyen

CRC Press
Taylor & Francis Group
Boca Raton London New York

CRC Press is an imprint of the
Taylor & Francis Group, an **informa** business
A CHAPMAN & HALL BOOK

Designed Cover Image: Nguyet Nguyen

First edition published 2026
by CRC Press
2385 NW Executive Center Drive, Suite 320, Boca Raton FL 33431

and by CRC Press
4 Park Square, Milton Park, Abingdon, Oxon, OX14 4RN

CRC Press is an imprint of Taylor & Francis Group, LLC

Library of Congress Cataloging-in-Publication Data
Names: Nguyen, Nguyet author
Title: Hidden Markov Models in Finance and Other Applications / Nguyet Nguyen.
Description: First edition. | Boca Raton : CRC Press, 2025. | Includes bibliographical references and index.
Identifiers: LCCN 2025029941 (print) | LCCN 2025029942 (ebook) | ISBN 9781041003717 hardback | ISBN 9781041003922 paperback | ISBN 9781003609599 ebook
Subjects: LCSH: Finance--Mathematical models | Markov processes
Classification: LCC HG106 .N5 2025 (print) | LCC HG106 (ebook)
LC record available at https://lccn.loc.gov/2025029941
LC ebook record available at https://lccn.loc.gov/2025029942

ISBN: 978-1-041-00371-7 (hbk)
ISBN: 978-1-041-00392-2 (pbk)
ISBN: 978-1-003-60959-9 (ebk)

DOI: 10.1201/9781003609599

Typeset in CMR10 font
by KnowledgeWorks Global Ltd.

Publisher's Note: This book has been prepared from camera-ready copy provided by the author.

Contents

Preface

This book aims to provide fundamental concepts and algorithms of the hidden Markov model (HMM) and its applications in finance such as stock price predictions and other areas such as speech recognition. Following the first application of HMM in speech recognition in 1966 [12], the applications of HMM have been extended to many areas. The wide application to many fields makes HMM very attractive to researchers in both academia and industry. Our target audience are those with basic knowledge in probability, statistics, and programming such as undergraduate students and young researchers. The readers of the book will learn the concepts and algorithms of the HMM through definitions, real-life examples, and R code. Researchers, or future researchers, can use the provided R code to explore the applications of the HMM with their own data. To familiarize readers with the HMM, a simple "Ball Game" example is used through many concepts and algorithms of the model in Chapters 1 and 2. Chapter 2 consists of practical examples which could be solved by a hand-held calculator or R code to help readers understand the model algorithms line by line. Chapter 3 presents some model selection methods. Chapter 4 presents applications of the HMM in finance that are based on our own research in the area. Chapter 5 summarizes HMM applications in other fields, such as speech recognition and bio-informatics.

1

Introduction

The hidden Markov model (HMM) is a regime-shift model that assumes observation data are driven by hidden regimes (or states). The model has been used in many fields, such as speech recognition, handwriting recognition, bioinformatics, and financial economics. In this chapter, we will introduce a brief history of the HMM, its main concepts, formulas to calculate the likelihood of the observations, examples to demonstrate the formulas, and some types of HMMs.

1.1 A Brief History of HMM

HMM is a signal detection model, whose mathematical foundations were developed in 1966 by Baum and Petrie [12]. The two main concepts of an HMM model are observations and hidden states arranged in sequences. The sequences of observations are visible, created by the sequence of hidden states. The hidden state sequence is a discrete sequence and satisfies the first order of a Markov process. The probability of moving from a state at time t to a state at time $t+1$ is assumed to depend only on the state at time t. This probability is called the transition probability. Figure 1.1 displays a sample of an HMM with three states. An observation at time t of an HMM is assumed to have a particular probability distribution corresponding to its hidden state. This probability distribution is called an observation distribution, or the emission distribution. Baum and Petrie [12] proved that with a given observation sequence, we can use the maximum likelihood estimate (MLE) to find the transition and emission probabilities. This process of finding unknown probabilities is called calibrating the parameters of the HMM.

Since its introduction in 1966, the HMM has undergone some major developments. These developments involved solving the problem of calibrating the model's parameters. The HMM has shifted from a model for a single observation sequence to a model for multiple observation sequences. In 1970, a maximization method was introduced in which the HMM parameters are calibrated using a single observation sequence in [13]. In the paper, the authors suggested using the model to predict weather and stock market behavior. Model training involves using the MLE algorithm, which includes $M - step$, where the model parameters are updated, and $E - step$, where the conditional expectation functions are calculated. Details about the $M - step$ and $E - step$ algorithms can be found in [85, 115]. In 1983, an MLE method was introduced to train the HMM with multiple observation sequences, assuming that all observation sequences are independent [86]. The authors [86] pointed out three problems of the Baum-Welch algorithm and presented a modified version of it. Then, they applied a particular class of the HMM model, called "left-to-right HMM", for speech recognition. In 1992, Leroux [84] proved that under some conditions, the parameters found by the MLE algorithm converge to the true parameters of the HMM with a probability of one. In 1996, Bickel and Ritov [17] proposed procedures to calibrate HMM parameters that are similar to

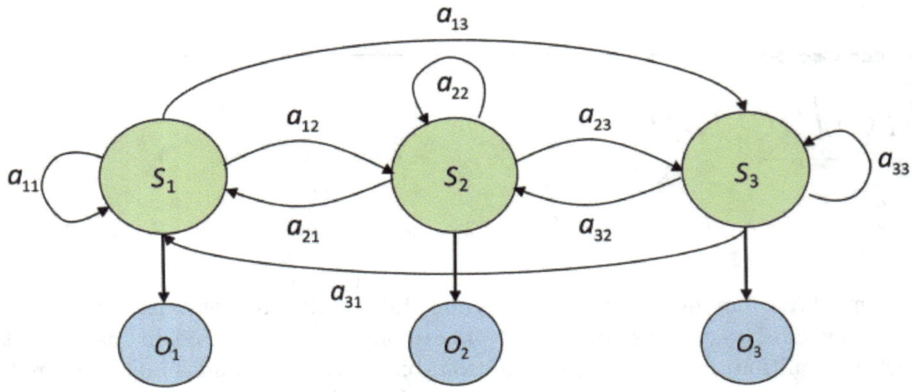

FIGURE 1.1
A hidden Markov model with a sequence of observations (O_1, O_2, and O_3) and a sequence of (hidden) states (S_1, S_2, and S_3). The transition probabilities between states i and j are a_{ij}.

the MLE algorithm used in [12], and proved that the procedures are asymptotically normal. In 1998, Bickel, Ritov, and Ryden [16] proved that under the weaker conditions that were used in [17], MLE is asymptotically normal.

In 2000, an HMM trained for multiple observation sequences was presented without the assumption of being independent of these sequences in [88]. In the paper, the authors also indicated that the HMM was trained for two special cases: independence observation sequences and uniform dependence observation sequences. In the same year, Gales [51] published a paper on cluster adaptive training of HMMs. In 2001 a modification of the Baum-Welch algorithm for multiple observation sequences was introduced in [8]. That same year, a complete tutorial of HMM and Bayesian Networks was published in [55]. The new overview of connections between the HMM and Bayesian Networks made the model applicable to multiple-state sequences, multi-scale representations and mixtures of discrete and continuous variables.

Recently, many researchers have been working on improving the MLE training procedures of continuous HMMs for speech recognition. In 2002, Woodland and Povey [149] presented a large scale framework for training of large vocabulary speech recognition systems based on Gaussian mixture HMMs. The authors showed that the scheme based on the linear interpolation of the large scale structure and MLE objective functions reduced the danger of over-training, and gave the lowest error rate. In 2004, Ben-Yishai and Burshtein [15] introduced a discriminative training algorithm, which is a modification of the Baum-Welch algorithm to train the HMM for speech recognition and word-spotting systems. Their experimental results showed that the modified algorithm was better in performance and computational complexity compared to the original Baum-Welch algorithm. Margin-based methods, which are mainly used in speech recognition, were proposed in [69,125,159] for training parameters of the HMM. Although the study of Sha and Saul [125] focused on a simpler and convex problem, a new method was proposed for discriminative learning of continuous HMM in the complex unlabeled setting that could be used for non-convex problems [35]. The authors proved the convergence of the algorithm and applied it on handwritten and speech digit recognition.

The HMM training algorithms have been improved to solve problems in other fields. In 2008, Sloin and Burshtein [126] used Support Vector Machines (SVMs) to improve MLE training procedures of both discrete and continuous HMMs. The authors combined the SVMs with the MLE algorithms to train a toy problem and an isolated noisy digit recognition task. The authors found a better result compared to the original MLE. In 2016, linear-time dynamic programming recursions, called k-segments, were introduced in [132] to improve the HMM's algorithms to fit larger data sets and more complex models. In 2021, Li, Lee, and Liao [87] proposed a modification of the Baum-Welch algorithm to train the HMM only for particular biological problems.

Details about the history and theory of HMM and its applications in a variety of fields can be found in [32,39,100,140]. In the next section, we present key concepts and examples of Markov processes and the HMM.

1.2 Markov Process

In this section, we will present the main concepts of the Markov process and examples. Recall that the hidden state sequence of an HMM is a discrete Markov chain process. Thus, we will first present the definition of a Markov process and examples of Markov processes, then we will present the main concepts of the HMM and examples in the following section.

1.2.1 Markov Process

The Markov chain (or Markov process) was introduced in 1906 by the Russian mathematician Andrey Andreyevich Markov [98]. He applied the new mathematical concept to analyze poetry and find an upcoming word of a sentence. A Markov process is a sequence of events that satisfy three conditions:

1. The number of possible outcomes, or states, of each event is finite.
2. The probability of the current event depends only on the outcome of the previous event.
3. The probabilities from states to states are constant over time.

Therefore, a Markov process is a *memory less* process. This Markov process is called the first-order Markov process with stationary or homogeneous transition probabilities. In this book, all Markov processes are first-order Markov processes with stationary transition probabilities. A Markov process of two or higher order is a Markov process in which the probability of the current state depends on two or more preceding states.

A Markov process will be defined by the following three variables:

1. Number of possible outcomes, or states, of each event N.
2. An $N \times N$ matrix for the probability of state-to-state transitions, denoted by matrix $\mathbf{A}$.
3. The initial probability of being a state of the first event, denoted by vector $\mathbf{p}$, or $\mathbf{p}_1$, of length N.

There are two types of Markov processes including discrete-time and continuous-time Markov chains. In the former, the transitions between states occur at discrete times, while, in the latter, the transitions between states occur at random and continuous time points.

We denote the possible N states as $\{S_1, S_2, ..., S_N\}$, and the actual state at time t of an event in the Markov process as q_t. A discrete first-order Markov process will satisfy the following condition at any time t,

$$\begin{aligned} \mathbb{P}(q_t &= S_j | q_{t-1} = S_i, q_{t-2} = S_k, ...) \\ &= \mathbb{P}(q_t = S_j | q_{t-1} = S_i). \end{aligned} \tag{1.1}$$

A continuous first order Markov process will satisfy the condition for any time $t, s \geq 0$,

$$\begin{aligned} \mathbb{P}(q_{s+t} &= S_j | q_s = S_i, q_k = S_k, 0 \leq k < s) \\ &= \mathbb{P}(q_{s+t} = S_j | q_s = S_i), \end{aligned} \tag{1.2}$$

where t is the time that the event will stay in the state i before jumping to the state j.

In the definition of the Markov chain, the transition probability from state S_i to state S_j is independent from t. We denote

$$a_{ij} = \mathbb{P}(q_t = S_j | q_{t-1} = S_i), i, j = 1, 2, .., N \tag{1.3}$$

for a discrete Markov chain and

$$a_{ij} = \mathbb{P}(q_{t+s} = S_j | q_t = S_i), i, j = 1, 2, .., N \tag{1.4}$$

for a continuous Markov chain. The matrix $\mathbf{A} = (a_{ij})$ is a constant transition matrix of the Markov process which satisfies the conditions

$$a_{ij} \geq 0, \tag{1.5}$$

$$\sum_{j=1}^{N} a_{ij} = 1, i, j = 1, 2, .., N. \tag{1.6}$$

We will present examples of the discrete time and continuous time of Markov chains in the next sections.

1.2.2 Example of a Discrete Markov Process

If we assume that the weather condition of each day is one of the three possible states, namely, Sunny, Rainy, or Cloudy, then the number of states is $N = 3$. We denote the states by $S_1 =$ Sunny, $S_2 =$ Rainy, and $S_3 =$ Cloudy. We assume that the probability of moving from the current day to the next day of weather states is a constant matrix $\mathbf{A}$,

	Sunny	Rainy	Cloudy
Sunny	0.4	0.3	0.3
Rainy	0.2	0.5	0.3
Cloudy	0.6	0.2	0.2

(1.7)

or

$$\mathbf{A} = \begin{bmatrix} 0.4 & 0.3 & 0.3 \\ 0.2 & 0.5 & 0.3 \\ 0.6 & 0.2 & 0.2 \end{bmatrix}. \tag{1.8}$$

The HMM with three states $S_1 =$ Sunny, $S_2 =$ Rainy, and $S_3 =$ Cloudy is presented in Figure 1.2. Suppose that the first observed day is Sunny, then the initial probability of states for the first observed day is

$$\mathbf{p}_1 = \begin{bmatrix} 1, 0, 0 \end{bmatrix}. \tag{1.9}$$

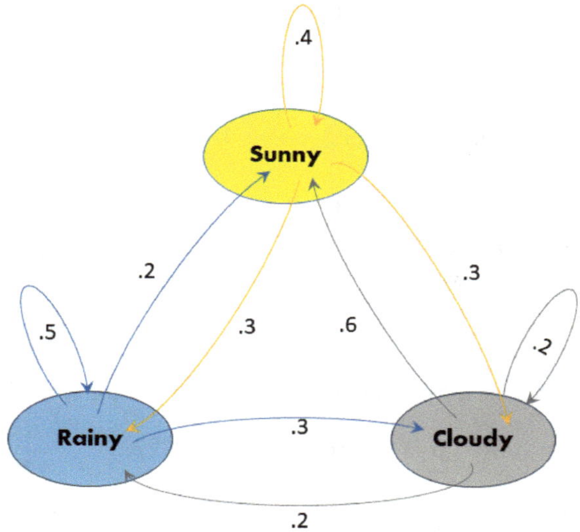

FIGURE 1.2
Diagram of transition probabilities between three weather states: Sunny, Rainy, and Cloudy.

Going forward, observations of weather conditions (or states) in the next days follow a Markov process. Number of states N, matrix $\mathbf{A}$, and the vector of initial probability $\mathbf{p}_1$ are parameters of the Markov process. With the parameters, we can calculate the probability of weather conditions in the second day, vector $\mathbf{p}_2$ by

$$\mathbf{p}_2 = \mathbf{p}_1 \cdot \mathbf{A} = \begin{bmatrix}1, 0, 0\end{bmatrix} \begin{bmatrix} 0.4 & 0.3 & 0.3 \\ 0.2 & 0.5 & 0.3 \\ 0.6 & 0.2 & 0.2 \end{bmatrix} = \begin{bmatrix}0.4, 0.3, 0.3\end{bmatrix}.$$

Thus, the probability of Sunny for the second day is 0.4, for Rainy it is 0.3, and for Cloudy it is 0.3. We can continue to calculate the probability of the weather condition for day three as

$$\mathbf{p}_3 = \mathbf{p}_2 \cdot \mathbf{A} = \begin{bmatrix}0.4, 0.3, 0.3\end{bmatrix} \begin{bmatrix} 0.4 & 0.3 & 0.3 \\ 0.2 & 0.5 & 0.3 \\ 0.6 & 0.2 & 0.2 \end{bmatrix} = \begin{bmatrix}0.4, 0.33, 0.27\end{bmatrix}.$$

The probability $\mathbf{p}_3$ can also be written as

$$\mathbf{p}_3 = \mathbf{p}_2 \cdot \mathbf{A} = (\mathbf{p}_1 \cdot \mathbf{A}) \cdot \mathbf{A} = \mathbf{p}_1 \cdot \mathbf{A}^2. \tag{1.10}$$

Then, we can calculate probability of weather condition for day n as

$$\mathbf{p}_n = \mathbf{p}_1 \cdot \mathbf{A}^{n-1}. \tag{1.11}$$

The prediction for a long period of time will be less accurate since the transition matrix A was trained by data points that are too far from the predicted point. Therefore, in practice, we should choose an appropriate length of the prediction period and then update the Markov process after each time period.

If the transition matrix $\mathbf{A}$ is a regular matrix, which means there is power of $\mathbf{A}$ that contains all positive non-zero entries, the probability $\mathbf{A}^n$ will converge to a vector $\mathbf{p}$, named the steady-state vector, when n goes to infinity.

$$\mathbf{p} = \lim_{n\to\infty} \mathbf{A}^n. \tag{1.12}$$

The steady-state vector is unique, independent from the initial vector $\mathbf{p}_1$, and has the property

$$\mathbf{p} \cdot \mathbf{A} = \mathbf{p}. \tag{1.13}$$

Therefore, we can find the steady vector of the weather example by solving the equation (1.13). From (1.13) we have

$$0 = \mathbf{p} \cdot \mathbf{A} - \mathbf{p} = \mathbf{p} \cdot \mathbf{A} - \mathbf{p} \cdot \mathbf{I} = \mathbf{p} \cdot (\mathbf{A} - \mathbf{I}),$$

where I is the identity matrix. Solving the system of equations, we have the result

$$\mathbf{p} = (0.3864, 0.3409, 0.2727).$$

We can also find p by taking the limit of the A^n. The limit of the power of the transition matrix A can be estimated by the following R code.

```
  A -> rbind(c(0.4,0.3,0.3),c(0.2,0.5,0.3),c(0.6,0.2,0.2))
  n -> 10
  for (i in 1:n){
      A=A %*% A
  print(A)
}
```

Running the code for $n = 10$, we have

$$\mathbf{A}^{10} = \begin{bmatrix} 0.3864 & 0.3409 & 0.2727 \\ 0.3864 & 0.3409 & 0.2727 \\ 0.3863 & 0.3409 & 0.2727 \end{bmatrix} \tag{1.14}$$

We see that as n becomes large, all rows of the matrix $\mathbf{A}^n$ converge to the steady vector $\mathbf{p}$, as shown in equation (1.12). That steady probability means that in the long run, on any n^{th} day, we will have 38.64% chance of being Sunny, 34.09% chance of being Rainy, and 27.27% chance of being Cloudy. Another question that we ask is: "If today is Rainy, based on the Markov chain model, what is the probability that the weather for the next five days will be Cloudy, Sunny, Sunny, Rainy, Cloudy?" Let's denote the sequence of weather observations in the six days by

$$O = \{S_2, S_3, S_1, S_1, S_2, S_3\}.$$

The initial probability at day 1 is $\mathbf{p} = (0, 0, 1)$. The probability of the observations is

$$\begin{aligned} \mathbb{P}(O|Model) &= \mathbb{P}(S_2)\mathbb{P}(S_3|S_2)\mathbb{P}(S_1|S_3)\mathbb{P}(S_1|S_1)\mathbb{P}(S_2|S_1)\mathbb{P}(S_3|S_2) \\ &= 1a_{23}a_{31}a_{11}a_{12}a_{23} \\ &= 1(0.3)(0.6)(0.4)(0.3)(0.3) \\ &= 0.00648. \end{aligned}$$

Using the Markov chain model, we can calculate the probability of any weather observation sequence. However, as the sequence gets longer, there is a higher chance that the probability of the observation sequence will approach zero due to the multiplication of multiple positive numbers less than one. We will discuss more about the issues with the HMM in the next chapter.

One more interesting question we may ask in the weather example is: "If the weather state in current day is S_i, what is the probability that it will stay in the same state for x days before changing to another state?". To answer the question, we need to calculate the probability of the observation sequences

$$O^{(j)} = \{\underbrace{S_i, S_i, ..., S_i}_{\text{x times}}, S_j, j \neq i\}. \tag{1.15}$$

The conditional probability of all of the observation sequence $O^{(j)}, j \neq i$ is:

$$\sum_{j \neq i} \mathbb{P}(O^{(j)}|Model, q_1 = S_i) = (a_{ii})^{x-1} \cdot (1 - a_{ii}). \tag{1.16}$$

Using equation (1.16), we can calculate the expectation of how many days the weather state will stay the same as the first day before changing to another state.

$$\begin{aligned} d &= \sum_{x=1}^{\infty} x(a_{ii})^{x-1}(1 - a_{ii}) \\ &= (1 - a_{ii}) \sum_{x=1}^{\infty} x(a_{ii})^{x-1} \\ &= (1 - a_{ii}) \frac{1}{(1 - a_{ii})^2} \\ &= \frac{1}{1 - a_{ii}}. \end{aligned}$$

Thus, based on the model, on average, the number of consecutive days of Sunny is $\frac{1}{1-0.4} \sim 1.7$ days, of Rainy is $\frac{1}{1-0.5} = 2$ days, and of Cloudy is $\frac{1}{1-0.2} = 1.25$ days.

The hidden state of an HMM is a discrete Markov process, so we will not present an example of continuous Markov process in the book. In the next sections, we will introduce the HMMs and their examples.

1.3 Hidden Markov Model

The HMM was defined based on the assumption that the observation data was driven by a hidden Markov process. Furthermore, the HMM is a stochastic signal model based on the following assumptions:

1. An observation at time t was generated by a hidden state.
2. The hidden states are finite and satisfy the first-order Markov property.
3. The matrix of transition probabilities between these states is constant.
4. The observation at time t of an HMM has a certain probability distribution corresponding with a possible hidden state.

From the definition of the HMM, we see that the HMM adds one more layer of probability: the probability that an observation corresponds to a certain hidden state, compared to the Markov process. There are two main types of HMMs: a discrete HMM and a continuous HMM. The two versions have minor differences, so we will first present key concepts of a discrete HMM. Then, we will add details about a continuous HMM.

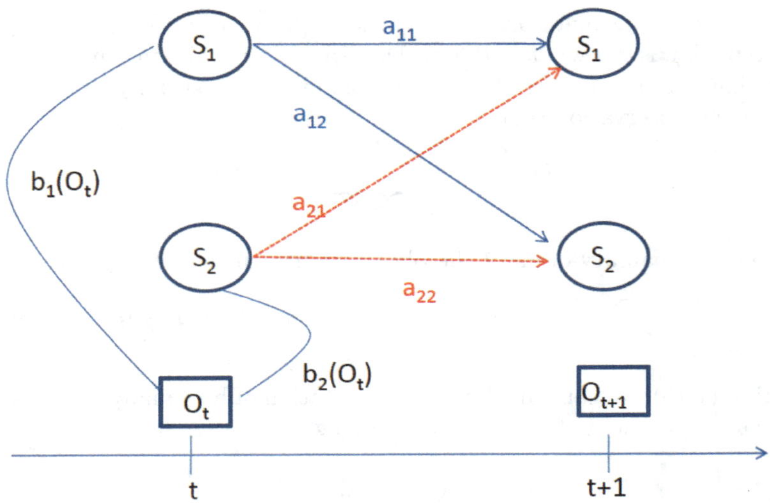

FIGURE 1.3
HMM with two states.

1.3.1 Discrete HMM

The basic elements of a finite discrete HMM are:

- Length of observation data, T
- Number of states, N
- Number of symbols per state, M
- Observation sequence, $O = \{O_t, t = 1, 2, \ldots, T\}$
- Hidden state sequence, $Q = \{q_t, t = 1, 2, \ldots, T\}$
- Possible values of each state, $S = \{S_i, i = 1, 2, \ldots, N\}$
- Possible symbols per state, $V = \{\nu_k, k = 1, 2, \ldots, M\}$
- Transition matrix, $\mathbf{A} = (a_{ij})$, where $a_{ij} = \mathbb{P}(q_t = S_j | q_{t-1} = S_i)$, and $i, j = 1, 2, \ldots, N$.
- Vector of initial probability of being in state (regime) S_i at time $t = 1$, $\mathbf{p} = (p_i)$, where $p_i = \mathbb{P}(q_1 = S_i)$, and $i = 1, 2, \ldots, N$
- Observation emission probability matrix, $\mathbf{B} = (b_{ik})$, where $b_{ik} = \mathbb{P}(O_t = v_k | q_t = S_i)$, $i = 1, 2, ..., N$ and $k = 1, 2, \ldots, M$.

A simple HMM with two states is presented in Figure 1.3. Note that the observation data O can be a single sequence or multiple sequences. For multiple observation sequences, we have the notations:

$$O = \{O^{(1)}, O^{(2)}, \ldots, O^{(L)}\},$$

where

$$O^{(l)} = \{O_1^{(l)}, O_2^{(l)}, \ldots, O_{T_l}^{(l)}\},$$

for $l = 1, \ldots, L$. Transition matrix $\mathbf{A}$, vector of initial probability $\mathbf{p}$ and observation probability $\mathbf{B}$ must satisfy the probability constraints:

$$a_{ij} \geq 0, \sum_{j=1}^{N} a_{ij} = 1, 1 \leq i, j \leq N. \tag{1.17}$$

$$b_{ik} > 0, \sum_{k=1}^{M} b_i(k) = 1, 1 \leq j \leq N, 1 \leq k \leq M. \tag{1.18}$$

$$p_i > 0, \sum_{i=1}^{N} p_i = 1, 1 \leq i \leq N. \tag{1.19}$$

In summary, the parameters of an HMM are the matrices $\mathbf{A}$ and $\mathbf{B}$ and the vector $\mathbf{p}$. For convenience, we use a compact notation for the parameters, given by:

$$\lambda \equiv \{\mathbf{A}, \mathbf{B}, \mathbf{p}\}. \tag{1.20}$$

1.3.2 Continuous HMM

In a continuous HMM, we will have infinite and continuous symbols for each hidden state. Therefore, the symbol v_k will be omitted from the model, and the emission probability of observation b_{ik} is modeled using a continuous probability distribution, and can be written as:

$$b_{ik} = b_i(O_t) = \mathbb{P}(O_t|q_t = S_i) = f(O_t|q_t = S_i), 1 \leq i \leq N, \tag{1.21}$$

where f is the density function of one or more distributions. The density function f is assumed to have ellipsoidal symmetry (Liporace, 1982) so that we can re-estimate the HMM parameters using the Baum-Welch algorithm. The algorithm will be discussed in the next chapter. In the paper, Liporace also pointed out that the Baum-Welch algorithm experiences some difficulties when the densities $b_i(O_t)$ are not log-concave. The author introduced a form of the density function:

$$b(y) = \int_0^{\infty} \mathcal{N}(y, \mu, \nu^2\Sigma) d\mathcal{G}(\nu), \tag{1.22}$$

where $\mathcal{G}$ is a probability function on $(0, \infty)$, and $\mathcal{N}(y, \mu, \nu^2\Sigma)$ is the multivariate Gaussian density with mean vector μ and covariance matrix $\nu^2\Sigma$. He proved that if an HMM satisfies the two conditions:

1. For each state S_i, the emission probability of observation $b_i(O_t)$ has the form (1.22),
2. Among $d + 1$ observation points, from the observation data, any d of which are linearly independent,

then the re-estimated of the HMM parameters process in the Baum-Welch algorithm will converge to a unique parameter λ^*. Common uses of the emission distributions are Gaussian distributions or a mix of the distributions. We present a continuous HMM where we assume that the emission probability of observations is a $M-$ mixture Gaussian distribution. In this case, the probability of observation O_t at given state S_i is

$$b_i(O_t) = \sum_{m=1}^{M} c_{im} \mathcal{N}(O_t, \mu_{im}, \sigma_{im}), 1 \leq i \leq N, \tag{1.23}$$

where μ_{im} and σ_{im} are the mean and variance of the distribution corresponding to the state $S_i, 1 \leq i \leq N$, respectively, c_{im} is the mixture coefficient of the mth mixture in state S_i, and $\mathcal{N}$ is a Gaussian density function. Precisely,

$$b_i(O_t) = \sum_{m=1}^{M} c_{im} b_i(O_t|\mu_{im}, \sigma_{im}) = \sum_{m=1}^{M} c_{im} \frac{1}{2\pi\sigma_{im}} e^{-\frac{(O_t-\mu_{im})^2}{2\sigma_{im}^2}}. \quad (1.24)$$

The coefficients c_{im} must satisfy the stochastic constraint

$$c_{im} \geq 0, 1 \leq i \leq N, 1 \leq m \leq M \quad (1.25)$$

$$\sum_{m=1}^{M} c_{im} = 1, 1 \leq i \leq N. \quad (1.26)$$

Thus, the emission probability function $b_i(O_t)$ will satisfy the condition

$$\int_{-\infty}^{\infty} b_i(x)dx = 1, 1 \leq i \leq N. \quad (1.27)$$

The parameters of the HMM become:

$$\lambda \equiv \{\mathbf{A}, \mu, \sigma, \mathbf{c}, \mathbf{p}\}, \quad (1.28)$$

where μ and σ are the means and variances of the Gaussian distributions, respectively, and $\mathbf{c}$ is the mixture coefficient vector. The parameters of the HMM in the discrete form (1.20) or continuous form (1.28) will well define the HMM if they satisfy the model conditions (1.17)-(1.19).

In summary, to define an HMM, first we have to choose a number of states, N, and a number of symbols, M, per state, and specify the symbols. Then, establish the transition probability matrix A, the observation probability matrix $\mathbf{B}$, and the initial probability vector $\mathbf{p}$.

1.3.3 Likelihood of an HMM

Given the observations O and the parameter λ of an HMM, the likelihood of the HMM is the probability of observation given the model parameter, denoted by

$$\mathbb{P}(O|\lambda).$$

For a single observation sequence $O = \{O_1, O_2, ..., O_T\}$, the likelihood of the model can be calculated as the sum of the probabilities of the observations at all possible state sequences Q,

$$\mathbb{P}(O|\lambda) = \mathbb{P}(O_1, O_2, ..., O_T|\lambda) \quad (1.29)$$

$$= \sum_{All\ Q} \mathbb{P}(O|Q, \lambda)\mathbb{P}(\mathbf{Q}|\lambda). \quad (1.30)$$

For a state sequence

$$Q = \{q_1, q_2, ..., q_T\}, \quad (1.31)$$

where q_1 is the initial state at time $t = 1$, the probability of the observation sequence O given the state sequence Q is

$$\mathbb{P}(O|Q, \lambda) = \prod_{\mathbf{t=1}}^{\mathbf{T}} \mathbb{P}(\mathbf{O_t}|\mathbf{q_t}, \lambda). \quad (1.32)$$

In the HMM we assumed statistical independence of observations. Thus, we have

$$\mathbb{P}(O|Q,\lambda) = \mathbf{b_{q_1}(O_1)b_{q_2}(O_2)...b_{q_T}(O_T)}. \tag{1.33}$$

The probability of the state sequence Q given the model parameter λ is

$$\mathbb{P}(Q|\lambda) = \mathbf{p_{q_1}a_{q_1q_2}a_{q_2q_3}...a_{q_{T-1}q_T}}. \tag{1.34}$$

Therefore, the likelihood of the observation sequence is

$$\begin{aligned}\mathbb{P}(O|\lambda) &= \sum_{AllQ} \mathbb{P}(O|Q,\lambda)\mathbf{P(Q|\lambda)} & (1.35)\\ &= \sum_{AllQ} p_{q_1}b_{q_1}(O_1)a_{q_1q_2}b_{q_2}(O_2)...a_{q_{T-1}q_T}b_{q_T}(O_T). & (1.36)\end{aligned}$$

We see that by using the traditional equation (1.36), we need to do $2TN^T$ calculations to get the $\mathbb{P}(O|\lambda)$. There are N^T possible state sequences. Thus, we need to do $2T-1$ multiplications, and N^T additions as shown in the right hand side of the equation (1.36). When the number of observations T increases, the number of calculations grows exponentially even with the minimum number of states $N = 2$. For example if $N = 2, T = 100$, the number of calculations is $2(100)(2^{100}) \sim 2.5(10^{32})$. Thus, this method is inefficient when applied to a real problem requiring a large data set. We will discuss more efficient methods to calculate the likelihood of these observations in the next chapter.

For multiple observation sequences O, we use the notation

$$O = \{O^{(1)}, O^{(2)}, ..., O^{(L)}\}, 1 \leq l \leq L, \tag{1.37}$$

where $O^{(l)}$ is a sequences of observation length T_l,

$$O^{(l)} = \{O_1^{(l)}, O_2^{(l)}, ..., O_{T_l}^{(l)}\}.1 \leq l \leq L. \tag{1.38}$$

If we assume that the observation sequences are independent, the likelihood of the observations is

$$\mathbb{P}(O|\lambda) = \prod_{\mathbf{l=1}}^{\mathbf{L}} \mathbb{P}(\mathbf{O^{(l)}}|\lambda). \tag{1.39}$$

Given appropriate parameters, we can use the HMM to generate the observation sequence $O = \{O_1, O_2, ..., O_T\}$ that maximizes the probability of observation $\mathbb{P}(O|\lambda)$. The observation at time t, O_t, is one of the symbols in V that maximizes the probability of observation $\mathbb{P}(O_|\lambda)$. We can generate O_t by following these steps.

1. Choose the initial state at time $t = 1$, $q_1 = S_i$ by using the initial probability p.
2. Set $t = 1$.
3. Choose $O_t = \nu_k$ by the using observation probability in state S_i. i.e. $b_i(k)$, from the matrix $\mathbf{B}$.
4. Generate q_{t+1} by using the transition probability matrix $\mathbf{A}$.
5. Set $t = t + 1$, and repeat steps (3-4) until $t = T$.

We will present some examples of discrete HMMs to illustrate the principles of the model in the next section.

1.3.4 Examples of Discrete HMMs

In this section, we will present some simple examples of discrete HMMs with two states $N = 2$, and two symbols for each state $M = 2$. Let us consider the "Ball Game" problem. Suppose that we have two boxes, representing the two states S_1 and S_2,

$$S = \{S_1, S_2\} = \{\text{Box 1, Box 2}\}.$$

Box one has three red balls and one Blue ball. Box two has two red balls and three Blue balls. The Red ball and Blue ball are the two symbols of each state (or box):

$$V = \{\nu_1, \nu_2\} = \{\text{Red, Blue}\}.$$

A player will first choose one of the two boxes by using a random process (e.g. flipping a coin) and then select a ball from the chosen box with replacement. The player repeats the process T times and records the selected balls. In the "Ball Game" we will have:

- The sequence of selected balls is the observation data, for example,

$$O = \{\text{Red, Blue, Blue, Red,...}\}.$$

- The sequence of selected boxes is the hidden state, for example,

$$Q = \{S_1, S_2, S_2, S_1, ...\} = \{\text{Box 1, Box 2, Box 2, Box 1,..}\}.$$

We assume that the hidden states of the HMM satisfy the first-order Markov chain, which means given a state sequence $Q = \{q_1, q_2, ..., q_T\}$, the probability of being in state S_i at time $t + 1$ depends only on q_t:

$$\mathbb{P}(q_{t+1} = S_i | q_1, q_2, ..., q_t) = \mathbb{P}(q_{t+1} = S_i | q_t).$$

Then, we calculate the probability of getting a ball based on a given box, the probability of observation matrix **B**. Because we assumed that box one has three Red balls and one Blue ball, and box two has two Red balls and three Blue balls, we can calculate the probability of observation matrix **B**:

- If the state S_1 (or Box 1) was selected at time t:
  - The probability of observing a Red ball is

$$b_1(1) = \mathbb{P}(\nu_1 | S_1) = \mathbb{P}(\text{Red}|\text{Box 1}) = \frac{3}{4} = 0.75.$$

  - The probability of observing a Blue ball is

$$b_1(2) = \mathbb{P}(\nu_2 | S_1) = \mathbb{P}(\text{Blue}|\text{Box 1}) = \frac{1}{4} = 0.25.$$

- If the state S_2 (or box two) was selected at time t:
  - The probability of observing a Red ball is

$$b_2(1) = \mathbb{P}(\nu_1 | S_2) = \mathbb{P}(\text{Red}|\text{Box 2}) = \frac{2}{5} = 0.4.$$

  - The probability of observing a Blue ball is

$$b_2(2) = \mathbb{P}(\nu_2 | S_2) = \mathbb{P}(\text{Blue}|\text{Box 2}) = \frac{3}{5} = 0.6.$$

Thus, we have the probability of observation matrix

$$\mathbf{B} = \begin{bmatrix} 0.75 & 0.25 \\ 0.40 & 0.60 \end{bmatrix}. \tag{1.40}$$

We will present some examples of the HMM based on the "Ball Game". Therefore, the probability of observation, matrix **B**, will be used for the following examples.

Example 1.1 *Suppose that the player in the "Ball Game" needs to choose the boxes (states) first by flipping a fair coin, then randomly pick up a ball from the chosen box. If the player plays twice, $T = 2$, what is the initial probability* **b** *and the transition probability matrix of the HMM?*

Solution 1.1
Because the player flips a fair coin to choose the box, the probability of getting a head or tail for each flip is .5. Thus, the transition probability matrix **A** will be

$$\mathbf{A} = \begin{bmatrix} 0.5 & 0.5 \\ 0.5 & 0.5 \end{bmatrix}, \tag{1.41}$$

and the initial probability **b** is

$$\mathbf{p} = (0.5, 0.5).$$

Note that the vector of initial probability **p** can be assigned to any values that satisfy condition 1.19. However, in the examples in the section, we assign a specific state for the first observation at time $t = 1$, $\mathbf{p} = (1, 0, ..., 0)$, for convenience.

Now we will present other examples based on the "Ball Game" which will have a more practical transition probability matrix **A** and an initial probability **p**.

Example 1.2 (Find the likelihood of the observations) *In the "Ball Game", assume that a player chooses Box 1 for the first time, $t = 1$, and then randomly picks up a ball from the box. He chooses a box again from the two boxes by using the given transition matrix and continues the process. The transition matrix given to choose the boxes (states),* **A***, is*

$$\mathbf{A} = \begin{bmatrix} 0.7 & 0.3 \\ 0.2 & 0.8 \end{bmatrix}. \tag{1.42}$$

What is the probability that the player will get the observation sequence $O = \{Red, Blue, Blue\}$ after playing the game three times?

Solution 1.2
We will solve this problem by listing all possible hidden state sequences Q of the observation sequence $O = \{\text{Red, Blue, Blue}\}$, then calculating the likelihood $\mathbb{P}(O|\lambda)$ using equation (1.36). Recall that the observation probability is

$$\mathbf{B} = \begin{bmatrix} 0.75 & 0.25 \\ 0.40 & 0.60 \end{bmatrix}.$$

Since the state S_1 is chosen first, $\mathbf{p} = (1, 0)$, we will have four possible state sequences for the observation O:

$$\begin{aligned} Q_1 &= \{S_1, S_1, S_1\} = \{\text{Box 1, Box 1, Box 1}\} \\ Q_2 &= \{S_1, S_1, S_2\} = \{\text{Box 1, Box 1, Box 2}\} \\ Q_3 &= \{S_1, S_2, S_1\} = \{\text{Box 1, Box 2, Box 1}\} \\ Q_4 &= \{S_1, S_2, S_2\} = \{\text{Box 1, Box 2, Box 2}\}. \end{aligned}$$

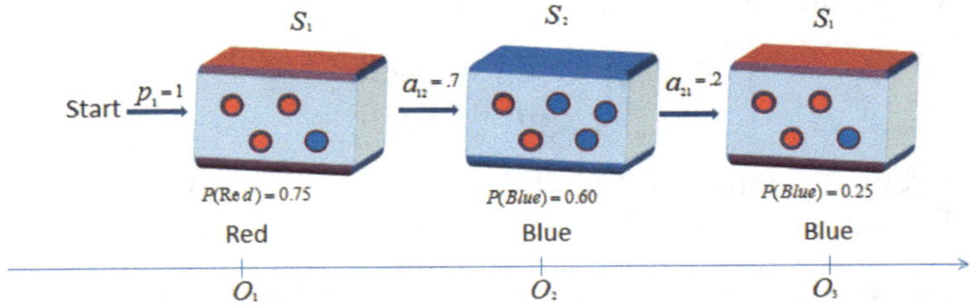

FIGURE 1.4
The "Ball Game" in Example 1.2 with state sequence $Q_3 = \{S_1, S_2, S_1\} = \{\text{Box 1, Box 2, Box 1}\}$.

Using the likelihood equation (1.36), we have:

$$\mathbb{P}(O|\lambda) = \sum_{\mathbf{l=1}}^{\mathbf{4}} \mathbb{P}(\mathbf{O}|\mathbf{Q_l}, \lambda) \cdot \mathbb{P}(\mathbf{Q_l}|\lambda). \tag{1.43}$$

Using equation (1.32), we have:

$$\begin{aligned}
\mathbb{P}(O|Q_1, \lambda) &= \mathbb{P}(\text{Red}|\text{Box 1}) \cdot \mathbb{P}(\text{Blue}|\text{Box 1}) \cdot \mathbb{P}(\text{Blue}|\text{Box 1}) \\
&= b_1(1)b_2(1)b_2(1) = b_{11}b_{21}b_{21} = 0.75(0.25)(0.25) = 0.046875. \\
\mathbb{P}(O|Q_2, \lambda) &= \mathbb{P}(\text{Red}|\text{Box 1}) \cdot \mathbb{P}(\text{Blue}|\text{Box 1}) \cdot \mathbb{P}(\text{Blue}|\text{Box 2}) \\
&= b_1(1)b_2(1)b_2(2) = b_{11}b_{21}b_{22} = 0.75(0.25)(0.6) = 0.1125. \\
\mathbb{P}(O|Q_3, \lambda) &= \mathbb{P}(\text{Red}|\text{Box 1}) \cdot \mathbb{P}(\text{Blue}|\text{Box 2}) \cdot \mathbb{P}(\text{Blue}|\text{Box 1}) \\
&= b_1(1)b_2(1)b_2(1) = b_{11}b_{22}b_{21} = 0.75(0.6)(0.25) = 0.1125. \\
\mathbb{P}(O|Q_4, \lambda) &= \mathbb{P}(\text{Red}|\text{Box 1}) \cdot \mathbb{P}(\text{Blue}|\text{Box 2}) \cdot \mathbb{P}(\text{Blue}|\text{Box 2}) \\
&= b_1(1)b_2(1)b_2(1) = b_{11}b_{22}b_{22} = 0.75(0.6)(0.6) = 0.27.
\end{aligned}$$

Using equation (1.33), we have

$$\begin{aligned}
\mathbb{P}(Q_1|\lambda) &= \mathbb{P}(\{\text{Box 1, Box 1, Box 1}\}|\lambda) = \mathbf{p_1 a_{11} a_{11}} = \mathbf{1(0.7)(0.7)} = \mathbf{.49}. \\
\mathbb{P}(Q_2|\lambda) &= \mathbb{P}(\{\text{Box 1, Box 1, Box 2}\}|\lambda) = \mathbf{p_1 a_{11} a_{12}} = \mathbf{1(0.7)(0.3)} = \mathbf{0.21}. \\
\mathbb{P}(Q_3|\lambda) &= \mathbb{P}(\{\text{Box 1, Box 2, Box 1}\}|\lambda) = \mathbf{p_1 a_{12} a_{21}} = \mathbf{1(0.3)(0.2)} = \mathbf{0.06}. \\
\mathbb{P}(Q_4|\lambda) &= \mathbb{P}(\{\text{Box 1, Box 2, Box 2}\}|\lambda) = \mathbf{p_1 a_{12} a_{22}} = \mathbf{1(0.3)(0.8)} = \mathbf{0.24}.
\end{aligned}$$

Finally, using the likelihood equation (1.43), we have

$$\mathbb{P}(O|\lambda) = (\mathbf{0.046875})(\mathbf{0.49}) + (\mathbf{0.1125})(\mathbf{0.21}) + (\mathbf{0.1125})(\mathbf{0.06}) + (\mathbf{0.27})(\mathbf{0.24}) = \mathbf{0.118144}.$$

Figure 1.4 displays the HMM Example 1.2 with state sequence Q_3.

Based on the result from Example 1.2, we can derive another "Ball Game" problem:

Example 1.3 (Find the hidden states) *In the "Ball Game", given that the initial probability* $\mathbf{p} = (1, 0)$, *the observation sequence* $O = \{Red, Blue, Blue\}$, *the transition matrix*

$$\mathbf{A} = \begin{bmatrix} 0.7 & 0.3 \\ 0.2 & 0.8 \end{bmatrix},$$

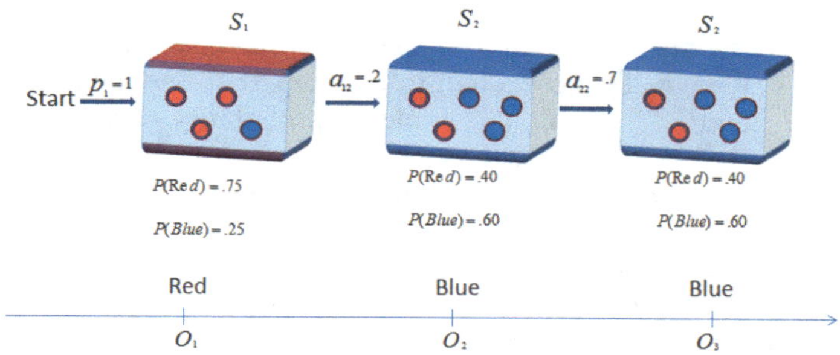

FIGURE 1.5
The "Ball Game" in Example 1.4 with state sequence $Q_4 = \{S_1, S_2, S_2\} = \{\text{Box 1, Box 2, Box 2}\}$.

and recall that the observation probability is

$$\mathbf{B} = \begin{bmatrix} 0.75 & 0.25 \\ 0.40 & 0.60 \end{bmatrix},$$

find the "best fit" hidden state sequence $Q = \{q_1, q_2, q_3\}$ *that maximizes the likelihood.*

Solution 1.3
We will find the hidden state sequence Q that maximizes the probability of the observation O. That means find the state sequence Q_k, which maximizes the likelihood of the observations.

$$k = \arg \max_{1 \le i \le 4} \{\mathbb{P}(O, Q_i|\lambda\} = \arg \max_{\mathbf{1 \le i \le 4}} \{\mathbb{P}(\mathbf{O}|\mathbf{Q_i}, \lambda) \cdot \mathbb{P}(\mathbf{Q_i}|\lambda)\}.$$

Based on the solution of the Example 1.2, we see that the state sequence Q_4 has the biggest observation probability among the four possible state sequences Q_1, Q_2, Q_3, Q_4. Using results of calculating $\mathbb{P}(O|Q_4, \lambda)$ and $\mathbb{P}(Q_4|\lambda$ from Example 1.2, we have

$$\mathbb{P}(O, Q_4|\lambda) = \mathbb{P}(\mathbf{O}|\mathbf{Q_4}, \lambda) \cdot \mathbb{P}(\mathbf{Q_4}|\lambda) = \mathbf{0.27(0.24)} = \mathbf{0.0648}.$$

Figure 1.5 represents result of the Example 1.4 with the optimal state sequence Q_4. Now we move on to the example that will demonstrate the process of choosing the observation O given the HMM parameters mentioned in (1.3.3).

Example 1.4 (Find optimal observation sequence) *In the "Ball Game", given that the initial probability* $\mathbf{p} = (1, 0)$, *the number of observations* $T = 3$, *the transition matrix*

$$\mathbf{A} = \begin{bmatrix} 0.7 & 0.3 \\ 0.2 & 0.8 \end{bmatrix},$$

and recall that the observation probability

$$\mathbf{B} = \begin{bmatrix} 0.75 & 0.25 \\ 0.40 & 0.60 \end{bmatrix},$$

find an optimal observation sequence $O = \{O_1, O_2, O_3\}$ *that maximizes the likelihood.*

Solution 1.4

We will follow the process given in (1.3.3) to find O. Because $\mathbf{p} = \begin{bmatrix}1, 0\end{bmatrix}$, at time $t = 1$ we have $q_1 = S_1 =$ Box 1. Next, we will choose a ball from Box 1 that has the maximum probability of success. In Box 1, the Red ball will have a higher probability of success. Thus, we choose the first ball as $O_1 =$ Red. Then, we will choose a box for time $t = 2$ given that Box 1 was chosen at time $t = 1$. Based on the transition matrix A, we have a bigger chance of moving from Box 1 to Box 1, $a_{11} > a_{12}$. Thus, we will choose Box 1 for time $t = 2$, $q_2 = Box1$. Again, from the Box 1 we should choose the Red ball because it has a higher probability of success. Thus, $O_2 =$ Red. Repeat the process for time $t = 3$, we also have $q_3 =$ Box 1, and $O_3 =$ Red. Thus, we have the observation sequence $O = \{$Red, Red, Red$\}$ and the corresponding hidden state sequence $Q = \{$Box 1, Box 1, Box 1$\}$. Applying equations(1.32) and (1.33), we have the probability of the observation

$$\mathbb{P}(O, Q|\lambda) = \mathbb{P}(\mathbf{O}|\mathbf{Q}, \lambda) \cdot \mathbb{P}(\mathbf{Q}|\lambda) = (\mathbf{b_{11}})^{\mathbf{3}}\mathbf{p_1}(\mathbf{a_{11}})^{\mathbf{2}} = (\mathbf{0.75})^{\mathbf{3}}(\mathbf{1})(\mathbf{0.7})^{\mathbf{2}} = \mathbf{0.20671875.}$$

We will introduce the hardest question related to the "Ball Game" in Example 1.5.

Example 1.5 (Find parameters for the HMM) *Suppose that we have two boxes of balls, each box will have a number of Red and Blue balls. A player will randomly choose a ball from a box with replacement. The player will repeat the process three times. Assume that the transition probability of choosing a box from time t to t* $+ 1$ *is* $\mathbf{A}$*, the probability of observation matrix is* $\mathbf{B}$*, and the initial probability of choosing the first box at time* $t = 1$ *is* $\mathbf{p}$ *are unknown. If the player chose the sequence of balls* $O = \{$*Red, Blue, Blue*$\}$*, and the process of choosing the balls based on an HMM with unknown parameter* $\lambda = \{\mathbf{A}, \mathbf{B}, \mathbf{p}\}$*, can you find the parameter?*

In order to answer this question, we have to find the set of model parameters, λ, that maximizes the probability of observation $\mathbb{P}(O|\lambda)$. Like other optimization strategies, we will start by guessing λ, calculate $\mathbb{P}(O|\lambda)$ then modify λ and repeat the steps until we locate the optimizer λ^* that gives the targeted observation probability. Since the process is too complicated to do by hand we will not present a complete solution for Example 1.5 here. Instead, we will describe the mathematical algorithm to solve this problem and demonstrate a full iteration of the training process in the next chapter.

1.3.5 Examples of a Continuous HMM

To demonstrate an example of a continuous HMM, we will return to the weather example of the discrete Markov chain in Section 1.2.2. Suppose that the observations of average daily temperature in a city in Florida are driven by the hidden Markov process in the weather example in Section 1.2.2 with the transition matrix

$$\mathbf{A} = \begin{bmatrix} 0.4 & 0.3 & 0.3 \\ 0.2 & 0.5 & 0.3 \\ 0.6 & 0.2 & 0.2, \end{bmatrix} \tag{1.44}$$

and the initial probability of being in the three states $S = \{$Sunny, Rainy, Cloudy$\}$ in the first day is $\mathbf{p} = \begin{bmatrix}1, 0, 0\end{bmatrix}$. We assume that the average temperature at time t corresponding to state S_i will have a probability of Gaussian distribution with mean and standard deviation $(\mu_i, \sigma_i), i = 1, 2, 3$. Suppose that

$$\mu = (\mathbf{60}, \mathbf{55}, \mathbf{50})$$

and

$$\sigma = (\mathbf{3}, \mathbf{2}, \mathbf{2.5}).$$

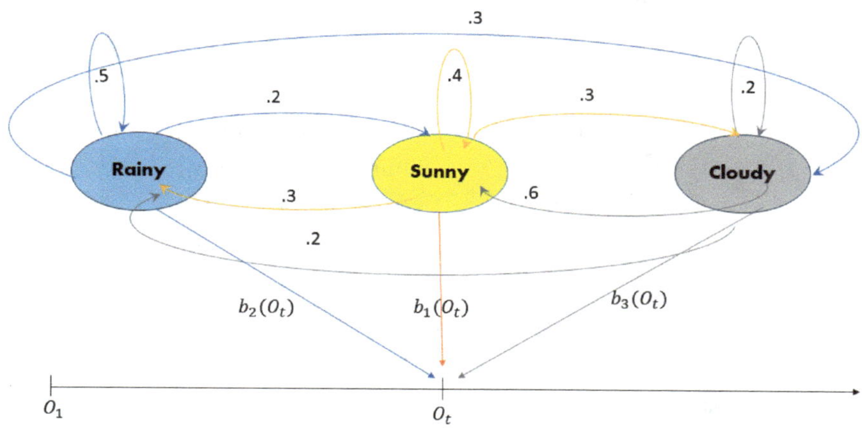

FIGURE 1.6
The continuous HMM for weather temperature with three hidden states: Sunny, Rainy, Cloudy.

The emission probability for the observation at time t corresponding to the state S_i is

$$\begin{aligned} b_i(O_t) &= b_i(O_t|\mu_i, \sigma_i) && (1.45) \\ &= \frac{1}{2\pi\sigma_i} e^{-\frac{(O_t-\mu_i)^2}{2\sigma_i^2}}, && (1.46) \end{aligned}$$

for $i = 1, 2, 3$. Figure 1.6 displays the continuous HMM for average daily temperature with three hidden states $S = \{\text{Sunny, Rainy, Cloudy}\}$.

Thus, the parameters of the HMM can be written in a compact notation:

$$\lambda = \{\mathbf{A}, \mu, \sigma, \mathbf{p}\}. \tag{1.47}$$

We can ask the question: what is the probability that the temperature of the first day is $O_1 = 61°F$? We have

$$\mathbb{P}(O_1|\lambda) = \mathbf{b_1(O_1)} = \frac{\mathbf{1}}{\mathbf{2\pi\sigma_1}} \mathbf{e}^{-\frac{(\mathbf{O_1}-\mu_1)^2}{2\sigma_1^2}} = \frac{\mathbf{1}}{\mathbf{2\pi 3}} \mathbf{e}^{-\frac{(\mathbf{61-60})^2}{\mathbf{2(3^2)}}} = \mathbf{0.1258}.$$

Using the density function of normal distribution, *dnorm*, in R we have

$$\mathbb{P}(O_1|\lambda) = \mathbf{b_1(O_1)} = \mathbf{dnorm(80, 70, 3)} = \mathbf{0.1258}. \tag{1.48}$$

We will use the *dnorm* function in R in the following examples. Now, we will answer similar questions that we asked with the discrete HMM in the "Ball Game" examples. The following examples are in regards to a continuous HMM.

Example 1.6 (Find the likelihood of observations) *Suppose that the average of daily temperatures in a city named "CoolSummer" is the HMM with the parameter λ in (1.47). What is the probability that the city will have the average daily temperature of $O = \{61°F, 57°F, 49°F\}$ in three days?*

Solution 1.6

Because the first day was assumed to be a Sunny day, we will have nine possible hidden state sequences:

$$\begin{aligned}
Q_1 &= \{S_1, S_1, S_1\} = \{\text{Sunny, Sunny, Sunny}\}\\
Q_2 &= \{S_1, S_1, S_2\} = \{\text{Sunny, Sunny, Rainy}\}\\
Q_3 &= \{S_1, S_1, S_3\} = \{\text{Sunny, Sunny, Cloudy}\}\\
Q_4 &= \{S_1, S_2, S_1\} = \{\text{Sunny, Rainy, Sunny}\}\\
Q_5 &= \{S_1, S_2, S_2\} = \{\text{Sunny, Rainy, Rainy}\}\\
Q_6 &= \{S_1, S_2, S_3\} = \{\text{Sunny, Rainy, Cloudy}\}\\
Q_7 &= \{S_1, S_3, S_1\} = \{\text{Sunny, Cloudy, Sunny}\}\\
Q_8 &= \{S_1, S_3, S_2\} = \{\text{Sunny, Cloudy, Rainy}\}\\
Q_9 &= \{S_1, S_3, S_3\} = \{\text{Sunny, Cloudy, Cloudy}\}.
\end{aligned}$$

Then using the likelihood equation (1.36), we have

$$\mathbb{P}(O|\lambda) = \sum_{\mathbf{l=1}}^{\mathbf{9}} \mathbb{P}(\mathbf{O}|\mathbf{Q_l}, \lambda) \cdot \mathbb{P}(\mathbf{Q_l}|\lambda). \qquad (1.49)$$

Using equation (1.32), we have:

$$\begin{aligned}
\mathbb{P}(O|Q_1, \lambda) &= dnorm(61, 60, 3)dnorm(57, 60, 3)dnorm(49, 60, 3) = 0.000002\\
\mathbb{P}(O|Q_2, \lambda) &= dnorm(61, 60, 3)dnorm(57, 60, 3)dnorm(49, 55, 2) = 0.000024\\
\mathbb{P}(O|Q_3, \lambda) &= dnorm(61, 60, 3)dnorm(57, 60, 3)dnorm(49, 50, 2.5) = 0.001495\\
\mathbb{P}(O|Q_4, \lambda) &= dnorm(61, 60, 3)dnorm(57, 55, 2)dnorm(49, 60, 3) = 0.000002\\
\mathbb{P}(O|Q_5, \lambda) &= dnorm(61, 60, 3)dnorm(57, 55, 2)dnorm(49, 55, 2) = 0.000034\\
\mathbb{P}(O|Q_6, \lambda) &= dnorm(61, 60, 3)dnorm(57, 55, 2)dnorm(49, 50, 2.5) = 0.002242\\
\mathbb{P}(O|Q_7, \lambda) &= dnorm(61, 60, 3)dnorm(57, 50, 2.5)dnorm(49, 60, 3) = 0.000000\\
\mathbb{P}(O|Q_8, \lambda) &= dnorm(61, 60, 3)dnorm(57, 50, 2.5)dnorm(49, 55, 2) = 0.000001\\
\mathbb{P}(O|Q_9, \lambda) &= dnorm(61, 60, 3)dnorm(57, 50, 2.5)dnorm(49, 50, 2.5) = 0.000059.
\end{aligned}$$

Using equation (1.33), we have:

$$\begin{aligned}
\mathbb{P}(Q_1|\lambda) &= \mathbb{P}(\{S_1, S_1, S_1\}|\lambda) = \mathbf{p_1 a_{11} a_{11}} = \mathbf{1(0.4)(0.4)} = \mathbf{0.16}\\
\mathbb{P}(Q_2|\lambda) &= \mathbb{P}(\{S_1, S_1, S_2\}|\lambda) = \mathbf{p_1 a_{11} a_{12}} = \mathbf{1(0.4)(0.3)} = \mathbf{0.12}\\
\mathbb{P}(Q_3|\lambda) &= \mathbb{P}(\{S_1, S_1, S_3\}|\lambda) = \mathbf{p_1 a_{11} a_{13}} = \mathbf{1(0.4)(0.3)} = \mathbf{0.12}\\
\mathbb{P}(Q_4|\lambda) &= \mathbb{P}(\{S_1, S_2, S_1\}|\lambda) = \mathbf{p_1 a_{12} a_{21}} = \mathbf{1(0.3)(0.2)} = \mathbf{0.06}\\
\mathbb{P}(Q_5|\lambda) &= \mathbb{P}(\{S_1, S_2, S_2\}|\lambda) = \mathbf{p_1 a_{12} a_{22}} = \mathbf{1(0.3)(0.5)} = \mathbf{0.15}\\
\mathbb{P}(Q_6|\lambda) &= \mathbb{P}(\{S_1, S_2, S_3\}|\lambda) = \mathbf{p_1 a_{12} a_{23}} = \mathbf{1(0.3)(0.3)} = \mathbf{0.09}\\
\mathbb{P}(Q_7|\lambda) &= \mathbb{P}(\{S_1, S_3, S_1\}|\lambda) = \mathbf{p_1 a_{13} a_{31}} = \mathbf{1(0.3)(0.6)} = \mathbf{0.18}\\
\mathbb{P}(Q_8|\lambda) &= \mathbb{P}(\{S_1, S_3, S_2\}|\lambda) = \mathbf{p_1 a_{13} a_{32}} = \mathbf{1(0.3)(0.2)} = \mathbf{0.06}\\
\mathbb{P}(Q_9|\lambda) &= \mathbb{P}(\{S_1, S_3, S_3\}|\lambda) = \mathbf{p_1 a_{13} a_{33}} = \mathbf{1(0.3)(0.2)} = \mathbf{0.06}.
\end{aligned}$$

Finally, using the likelihood equation (1.49), we have

$$\begin{aligned}
\mathbb{P}(O|\lambda) &= (0.000002)(0.16) + (0.000024)(0.12) + (0.001495)(0.12)\\
&+ (0.000002)(0.06) + (0.000034)(0.15) + (0.002242)(0.09)\\
&+ (0.0)(0.18) + (0.000001)(0.06) + (0.000059)(0.06)\\
&= 0.0003932.
\end{aligned}$$

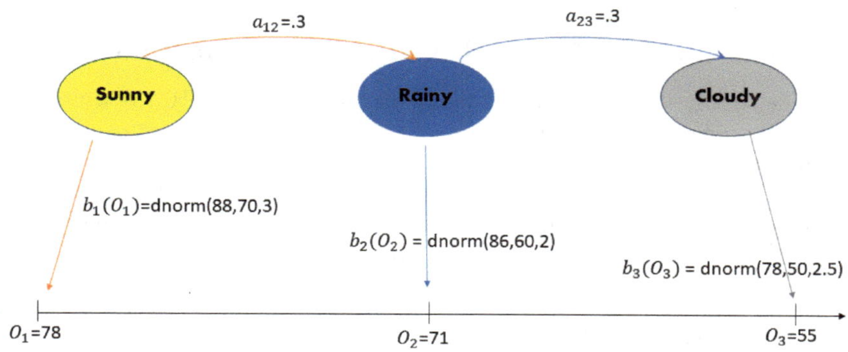

FIGURE 1.7
The continuous HMM observations $O = \{O_1, O_2, O_3\}$ with the corresponding hidden states: $Q_6 = \{Sunny, Rainy, Cloudy\}$.

Figure 1.7 represents the three observations with the corresponding hidden sates Q_6.

Example 1.7 (Find hidden states) *Suppose that the average of daily temperatures in Sarasota, Florida, USA is the HMM with the parameter λ in (1.47). Find the "best fit" hidden state sequence for the observations of average daily temperature in the city for three first days $O = \{61°F,\ 57°F,\ 49°F\}$.*

Solution 1.7
We need to find the hidden state sequence $Q = \{q_1, q_2, q_3\}$ that maximizes the probability of the observation O. That means find the state sequence Q_k, which maximizes the likelihood of the observations $\mathbb{P}(O, Q_k|\lambda)$, giving us

$$k = \arg\max_{1 \leq i \leq 9} \{\mathbb{P}(O, Q_i|\lambda\} = \arg\max_{\mathbf{1 \leq i \leq 9}} \{\mathbb{P}(\mathbf{O}|\mathbf{Q_i}, \lambda) \cdot \mathbb{P}(\mathbf{Q_i}|\lambda)\}.$$

From our solution to Example 1.6, we see that the state sequence $Q_6 = \{S_1, S_2, S_3\} = $ {Sunny, Rainy, Cloudy} has the biggest observation probability among the nine possible state sequences $Q_1, Q_2, ..., Q_9$. Therefore, the hidden state sequence is $Q = Q_6 = \{S_1, S_2, S_3\} = $ {Sunny, Rainy, Cloudy}.

We will present another example of the weather forecast problem without a solution because the example is too complicated to do it by hand. We will find solution for the example in the next chapter.

Example 1.8 (Find parameters for the HMM) *Given the daily average temperature in a city in one week in Fahrenheit measure:*

$$O = \{68°F,\ 64°F,\ 61°F,\ 60°F,\ 57°F,\ 52°F,\ 48°F\}$$

Suppose that the temperatures follow a continuous HMM with three hidden states. Can you find the parameters of the model?

1.4 Types of an HMM Based on Its Transition Matrix

In the previous sections, we presented the most commonly used HMM which has a hidden process given by a first-order Markov process. This HMM model is called the first-order HMM. In this section, we will present some types of HMMs based on the transition probabilities of the hidden states. Note that similar to the first-order HMM, each type of the HMMs has two versions: discrete and continuous based on the distribution of the conditional probability of the observations corresponding to their hidden states.

1.4.1 The Ergodic HMM

If every state of the HMM can be reached in finitely many steps from any other state of the model, the HMM is called the ergodic or fully connected HMM, (EHMM). The EHMM has the property that every state can be reached from any other states after a finite number of steps. The transition probabilities a_{ij} are positive for any $1 \leq i, j \leq N$. Figure 1.8 represents the transition probabilities between three states of an ergodic HMM, where all the probabilities are positive. The summary of the transition probabilities is on the Figure 1.9. Details about the EHMM and its algorithms can be found in [81, 117].

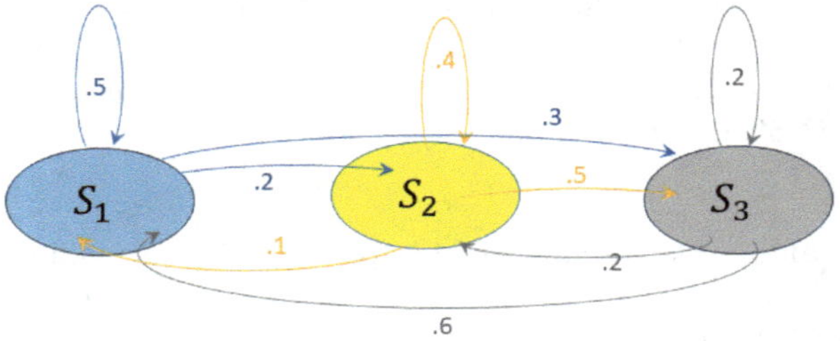

FIGURE 1.8
Transition probabilities between three states of an ergodic HMM.

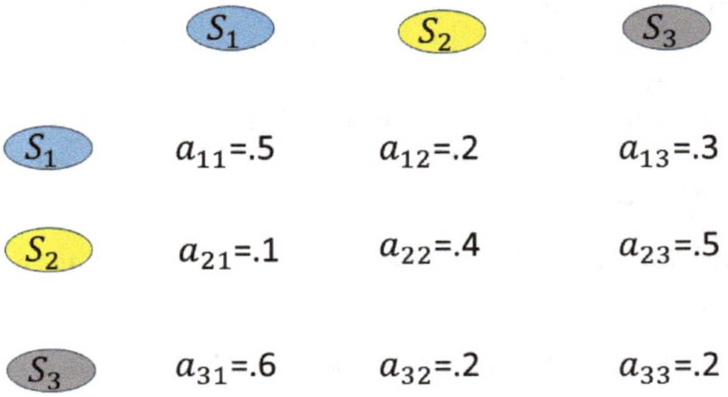

FIGURE 1.9
Transition probability matrix of the three states ergoric HMM.

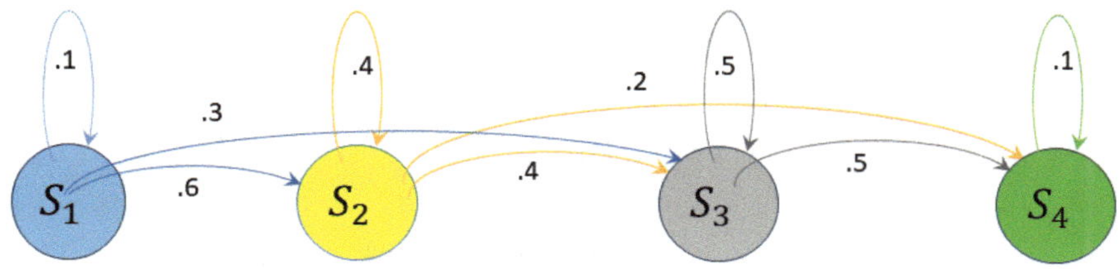

FIGURE 1.10
Transition probability matrix of the left-right four states HMM with $\Delta = 2$.

1.4.2 The Left-Right HMM

A left-right HMM, or a Bakis HMM, was proposed by Bakis [9]. The model better explained the behavior of the speech signals. In a left-right HMM, as time increases, the index of hidden state increases or stay the same. The transition probabilities of a left-right HMM have the properties

- No transitions are allowed to states whose indices are lower than the current state

$$a_{ij} = 0, j < i$$

- The initial state is S_1

$$p_i = \begin{cases} 1, i = 1 \\ 0, i \neq 1 \end{cases}$$

- No jump of more than Δ steps is allowed

$$a_{ij} = 0, j > i + \Delta, \Delta \geq 1.$$

Figures 1.10-1.11 represent an example of the left-right HMM with four states and the limited jumping state $\Delta = 2$.

1.4.3 Other Left-Right Model

Based on properties of an observation sequence and assumptions of its hidden Markov sequence, we may add some strict rules to the left-right model. For example we can add one more rule to the left-right model by not allowing a state with an odd index jump to states with even indices. Since these rules would change based on the properties and assumptions, many other variations of left-right models can exist.

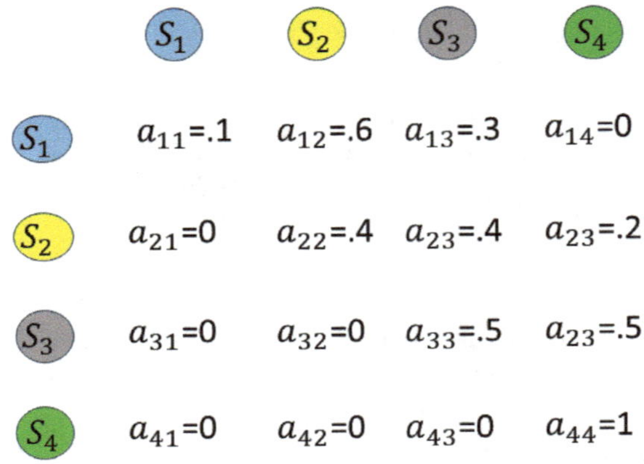

FIGURE 1.11
Transition probability between three states of the left-right HMM. with $\Delta = 2$.

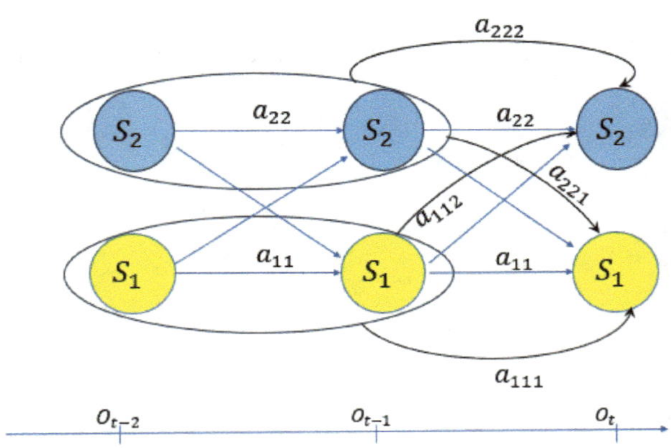

FIGURE 1.12
Transition probability matrix of the second-order two states HMM.

1.5 Higher-Order HMM

In a higher-order HMM, the current state depends not only on the previous state but also on a sequence of preceding states. For a second-order HMM, the transition matrix of the states is a three dimensional matrix $\mathbf{A} = (a_{ijk})$, where

$$a_{ijk} = \mathbb{P}(q_t = S_k | q_{t-1} = S_j, q_{t-2} = S_i),$$

$t = 2, ..., T$ and $1 \leq i, j, k \leq N$, where N is number of states and T is number of observations. For a second-order two states HMM, the matrix $A = (a_{ijk})$ will have eight entries. Some of the entries are presented in Figure 1.12.

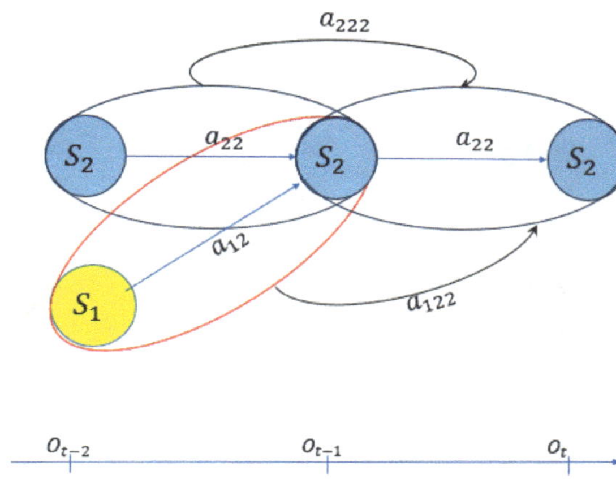

FIGURE 1.13
Transition probability matrix of the second-order two states HMM.

From Figure 1.13, we can see the relationship between the transition matrix of the first-order HMM and the second-order HMM. Researchers can promote algorithms for higher-order HMM by reducing the higher-order to the first-order HMM, then using the algorithms of the first-order HMM. Another way is to modify the algorithms of the first-order HMM to fit with the transition matrix of the higher-order HMM. An example of reducing the second-order HMM to the first order HMM is to form a new state sequence $\hat{q} = \{(q_{t-1}, q_t), t = 2, ...T\}$. Each of the entries in the new state sequence $\hat{q}$ is a pair of states (q_{t-1}, q_t). The probability transition a_{ijk} in the second-order HMM would be the probability transition a_{ij} to a_{jk} from state (q_{t-2}, q_{t-1}) to state (q_{t-1}, q_t) of the new sequence $\hat{q}$. The second-order state sequence q_t becomes the first-order sequence $\hat{q}$. Note that S_t has two states $\{S_1, S_2\}$, but $\hat{q}$ has four states $\{(S_1, S_1), (S_1, S_2), (S_2, S_1), (S_2, S_2)\}$. Figure 1.13 presents the two transitions of the new first-order sequence $\hat{q}$.

If we have the transition probabilities of the second-order HMM:

$$\begin{aligned} a &= a_{111} = \mathbb{P}(q_t = S_1 | q_{t-1} = S_1; q_{t-2} = S_1) \\ b &= a_{121} = \mathbb{P}(q_t = S_1 | q_{t-1} = S_2; q_{t-2} = S_1) \\ c &= a_{211} = \mathbb{P}(q_t = S_2 | q_{t-1} = S_1; q_{t-2} = S_1) \\ d &= a_{221} = \mathbb{P}(q_t = S_1 | q_{t-1} = S_2; q_{t-2} = S_2), \end{aligned}$$

then the first-order $\hat{q}$ with four states will have the transition probability matrix:

$$\mathbf{A} = \begin{bmatrix} a & 1-a & 0 & 0 \\ 0 & 0 & b & 1-b \\ c & 1-c & 0 & 0 \\ 0 & 0 & d & 1-d. \end{bmatrix}$$

Figure 1.14 represents the relationship between the second-order states and the first-order states.

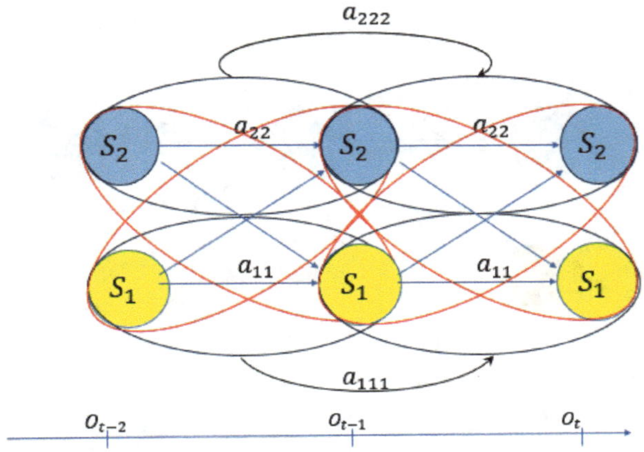

FIGURE 1.14
Transition probability matrix of the second-order two states HMM.

Details about high order HMMs and their algorithms can be found in [27,37,43,129,153, 157, 165]. Note that the rules of the transition matrix presented in the initial parameters of the first order HMM will remain throughout the reestimate parameters process in the Baum Welch algorithm that we will present in the next chapter. In this book, we focus on the fully first order HMM.

2

Three Main Problems of an HMM and Their Algorithms

In this chapter, we will present the three main problems that an HMM can solve: finding probability of observations, finding hidden states, and calibrating the model parameters. We will present HMM algorithms corresponding to these three problems and their applications to solving the examples of the "Ball Game" that were presented in Chapter 1. In practice, these algorithms have the problem of underflow when the observation data is large because the probability of the observations will be extremely small. Therefore, in this chapter, we will discuss the scaling method to fix this issue. We also present R code for each of the algorithms which can be applied to real data sets in the next chapters. R code of these algorithms and their applications are also presented in the chapter.

2.1 Three Main Problems of an HMM

In the "Ball Game" examples from the previous section, we mentioned the three questions, or problems, that we can solve when working with an HMM. In this section, we will present these three most significant problems of an HMM in a general case. When using an HMM to solve a real-world problem, we face three critical questions:

1. **Evaluation problem**: Given the observation data $O = \{O_t, t = 1, 2, \ldots, T\}$ and the model parameters $\lambda = \{\mathbf{A}, \mathbf{B}, \mathbf{p}\}$, compute the probabilities of the observations, $\mathbb{P}(O|\lambda)$.
2. **Decoding problem**: Given the observation data $O = \{O_t, t = 1, 2, \ldots, T\}$ and the model parameters $\lambda = \{\mathbf{A}, \mathbf{B}, \mathbf{p}\}$, find the best corresponding state sequence $Q = \{q_1, q_2, ..., q_T\}$ that produced O.
3. **Learning problem**: Given the observation data $O = \{O_t, t = 1, 2, \ldots, T\}$, find the model's parameters $\lambda = \{\mathbf{A}, \mathbf{B}, \mathbf{p}\}$ to maximize $\mathbb{P}(O|\lambda)$.

Researchers have found mathematical algorithms to solve each of these problems. The first were Baum and Egon [11] and Baum and Sell [14], who introduced the forward and backward algorithms to solve the first problem. Next, Viterbi [141] and Forney [47] developed the Viterbi algorithm to solve the second problem. Then, Baum, Petrie, Soules, and Weiss [13] presented the Baum-Welch algorithm to solve the third problem for a single observation sequence. Over the next three decades, Levinson, Rabiner, and Sondhi [86], and Li, Parizeau, and Plamondon [88] implemented algorithms based on the Baum-Welch algorithm to calibrate HMM parameters for multiple observation sequences. In the following sections, we introduce the forward algorithm, the backward algorithm, the Viterbi algorithm and the Baum-Welch algorithm. Either the forward or backward algorithm can be used to calculate the probability of observation in problem 1, while both of these algorithms are needed in the Baum-Welch algorithm. For convenience, we use the notation $b_i(O_t)$ for conditional probability $b_i(k)$.

2.2 Forward Algorithm

The forward algorithm is an algorithm used to find the probability of observation, $\mathbb{P}(O|\lambda)$, given the HMM parameters. In the "Ball Game" example 1.2 in Chapter 1, we solve this problem by listing all possible outcomes of the hidden states, calculating the probability of observations given each state sequence, and summing them up to find $\mathbb{P}(O|\lambda)$. However, this method is not efficient if we have a longer observation sequence since the process requires too many calculations. To counteract this issue, we can use the forward algorithm to calculate the probability of observation.

2.2.1 Forward Algorithm for a Single Observation Sequence

In the forward algorithm, we first calculate the likelihood of the model by considering the probability of observations if the last state is S_i, $\mathbb{P}(O_1, O_2, ..., O_t, q_T = S_i|\lambda)$, for $i = 1, ..., N$. Then, the observation probability $\mathbb{P}(O|\lambda)$ is the sum of these conditional probabilities:

$$\mathbb{P}(O|\lambda) = \sum_{\mathbf{i=1}}^{\mathbf{N}} \mathbb{P}(\mathbf{O_1, O_2, ..., O_t, q_T = S_i}|\lambda).$$

The probability $\mathbb{P}(O_1, O_2, ..., O_t, q_T = S_i|\lambda)$ can be calculated recursively by defining the forward probability function at time t as

$$\alpha_t(i) = \mathbb{P}(O_1, O_2, ..., O_t, q_t = S_i|\lambda), \mathbf{t = 1, ..., T, i = 1, ..., N}. \tag{2.1}$$

Figure 2.1 is the diagram of the forward probability function $\alpha_t(i)$.

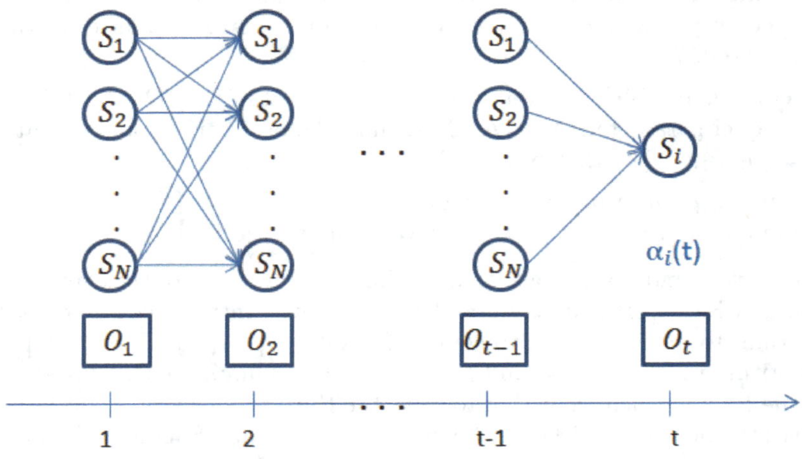

FIGURE 2.1
Diagram of the forward probability function $\alpha_t(i)$.

Recursively, we have the forward probability at time $t+1$ as

$$\alpha_{t+1}(j) = (\sum_{i=1}^{N} \alpha_t(i) a_{ij}) b_j(O_{t+1}), t = 1, .., T-1, j = 1, ..., N., \tag{2.2}$$

where a_{ij} and $b_j(O_{t+1})$ were defined in Section 1.3.2. Then, the probability of observation $\mathbb{P}(O|\lambda)$ is just the sum of $\alpha_T(i)$, for $i = 1, ..., N$,

$$\mathbb{P}(O|\lambda) = \sum_{\mathbf{i=1}}^{\mathbf{N}} \alpha_{\mathbf{T}}(\mathbf{i}).$$

Algorithm 1 presents the forward algorithm of the HMM for a single observation sequence.

Algorithm 1 The forward algorithm of the HMM with single observation sequence.

Initialization: for i = 1, 2, ..., N

$$\alpha_1(i) = p_i b_i(O_1).$$

Recursion: for $t = 2, 3, \ldots, T$, and for $j = 1, 2, \ldots, N$, compute

$$\alpha_t(j) = \left[\sum_{i=1}^{N} \alpha_{t-1}(i) a_{ij}\right] b_j(O_t).$$

Output:

$$\mathbb{P}(O|\lambda) = \sum_{\mathbf{i=1}}^{\mathbf{N}} \alpha_{\mathbf{T}}(\mathbf{i}).$$

2.2.2 Forward Algorithm for Multiple Observation Sequences

With minor changes from the forward algorithm for a single observation sequence in Section 2.2.1, we have the forward algorithm for multiple independent observation sequences $O = \{`O^{(1)}, O^{(2)}, ..., O^{(L)}\}$, where $O^{(l)} = \{O_1^l, O_2^l, ..., O_T^l\}$ for $l = 1, 2, ..., L$. We define the joint probability function as

$$\alpha_t^{(l)}(i) = \mathbb{P}(O_1^{(l)}, O_2^{(l)}, ..., O_t^{(l)}, q_t = S_i|\lambda), \mathbf{t = 1, 2, ..., T \text{ and } l = 1, 2, ..., L},$$

then we calculate $\alpha_t^{(l)}(i)$ recursively. The probability of observation $\mathbb{P}(O^{(l)}|\lambda)$ is just the sum of $\alpha_T^{(l)}(i)'s$. The algorithm 2 presents HMM forward algorithm multiple observation sequences.

2.2.3 Example of Using the Forward Algorithm

We can apply the forward algorithm to calculate the probability of observation, $\mathbb{P}(O|\lambda)$, in the Example 1.2 in Section 1.3.4 with two states and two symbols, $\nu_1 = Red$, and $\nu_2 = Blue$ for each state. Recall that the observation sequence is

$$O = \{O_1, O_2, O_3\} = \{Red, Blue, Blue\} = \{\nu_1, \nu_2, \nu_2\}.$$

Algorithm 2 The forward algorithm for multiple independent observation sequences.

Initialization $\mathbb{P}(O|\lambda) = \mathbf{1}$
For $l = 1, 2, ..., L$ do
Initialization: for i=1,2,..., N

$$\alpha_1^{(l)}(i) = p_i b_i(O_1^{(l)}).$$

Recursion: for $t = 2, 3, \ldots, T$, and for $j = 1, 2, \ldots, N$, compute

$$\alpha_t^{(l)}(j) = \left[\sum_{i=1}^{N} \alpha_{t-1}(i) a_{ij}\right] b_j(O_t^{(l)}).$$

Calculate:

$$P(O^{(l)}|\lambda) = \sum_{\mathbf{i=1}}^{\mathbf{N}} \alpha_{\mathbf{T}}^{(\mathbf{l})}(\mathbf{i}).$$

Update:

$$\mathbb{P}(O|\lambda) = \mathbb{P}(\mathbf{O}|\lambda) \cdot \mathbb{P}(\mathbf{O}^{(\mathbf{l})}|\lambda).$$

Output: $\mathbb{P}(O|\lambda)$.

The transition probability matrix, the observation probability matrix, and the initial probability of being in each state are respectively

$$\mathbf{A} = \begin{bmatrix} 0.7 & 0.3 \\ 0.2 & 0.8 \end{bmatrix}, \ \mathbf{B} = \begin{bmatrix} 0.75 & 0.25 \\ 0.40 & 0.60 \end{bmatrix}, \ \textbf{and } \mathbf{p} = [p_1, p_2] = [1, 0]. \tag{2.3}$$

The model parameter is $\lambda = \{\mathbf{A}, \mathbf{B}, \mathbf{p}\}$, number of states $N = 2$, $S_1 = Red$, $S_2 = Blue$, and number of observations $T = 3$. By using the forward probability function defined in 2.2, we have:

$$\mathbb{P}(O|\lambda) = \alpha_3(\mathbf{1}) + \alpha_3(\mathbf{2}) = \mathbf{P}(\mathbf{O_1}, \mathbf{O_2}, \mathbf{O_3}|\mathbf{q_3} = \mathbf{S_1}) + \mathbf{P}(\mathbf{O_1}, \mathbf{O_2}, \mathbf{O_3}|\mathbf{q_3} = \mathbf{S_2}),$$

where q_3 is the last state in the hidden state sequence $q = q_1, q_2, q_3$. Since $\mathbf{p} = (p_1, p_2) = (1, 0)$ at time t_1, we have:

$$\begin{aligned} \alpha_1(1) &= \mathbb{P}(O_1|q_1 = S_1) = \mathbb{P}(Red|q_1 = Red) \\ &= p_1 b_1(O_1) p_1 b_1(\nu_1) = p_1 b_{11} = (1)(0.75) \\ &= 0.75 \\ \alpha_1(2) &= \mathbb{P}(O_1|q_1 = S_2) = \mathbb{P}(Red|q_1 = Blue) \\ &= p_2 b_2(O_1) p_2 b_2(\nu_1) p_2 b_{21} = (0)(0.4) \\ &= 0. \end{aligned}$$

Using the recursive forward equation (2.2), at time t_2 we have:

$$\begin{aligned} \alpha_2(1) &= [\alpha_1(1)a_{11} + \alpha_1(2)a_{21}]b_1(O_2) \\ &= [0.75(0.7) + 0(.2)]b_1(\nu_2) = [0.75(0.7) + 0(.2)]b_{12} = [0.75(0.7) + 0(.2)]0.25 \\ &= 0.13125 \end{aligned}$$

$$\begin{aligned}\alpha_2(2) &= [\alpha_1(1)a_{12} + \alpha_1(2)a_{22}]b_2(O_2)\\ &= [0.75(0.3) + 0(.8)]b_2(\nu_2) = [0.75(0.3) + 0(.8)]b_{22} = [0.75(0.3) + 0(.8)](0.6)\\ &= 0.135.\end{aligned}$$

Similarly, at time t_3 we have

$$\begin{aligned}\alpha_3(1) &= [\alpha_2(1)a_{11} + \alpha_2(2)a_{21}]b_1(O_3)\\ &= [0.13125(0.7) + 0.135(0.2)](0.25)\\ &= 0.029719\\ \alpha_2(2) &= [\alpha_2(1)a_{12} + \alpha_2(2)a_{22}]b_2(O_3)\\ &= [0.13125(0.3) + 0.135(0.8)](0.6)\\ &= 0.088425.\end{aligned}$$

Thus, the probability of observation is

$$\mathbb{P}(O|\lambda) = \boldsymbol{\alpha_3(1) + \alpha_3(2) = 0.029719 + 0.088425 = 0.118144}.$$

This result is consistent with the result that we found in Example 1.2. However, using the recursive formula significantly reduces the number of calculations needed to find the probability of observation.

In general, with N observations and T states, the traditional method presented in Section 1.3.3 will need $2T(N^T)$ calculations. In contrast, using the forward algorithm presented in 1, we will only need N^2T calculations. In this example, using the traditional method in Section 1.3.3 we did $2(3)(2^3) = 48$ calculations, whereas using the forward algorithm 1, we only needed to do $2^2(3) = 12$ calculations. Therefore, as the number of observation T increases, using the forward algorithm 1 is more efficient when used to calculate the probability of observation. After the handheld demonstration, we will present the R code for the discrete HMM forward algorithm to calculate the probability of observations, and run the code using the data from Example 1.2.

2.2.4 R Code for the Forward Algorithm

Below is the R code using the forward algorithm to calculate the likelihood of observations of a discrete HMM:

```
# Dis.forward function calculates the likelihood of a discrete
# HMM using the forward algorithm.
# Input:
#    x : observation sequence (length n), translated to binary
#        values 1 or 2
#    A : transition matrix (m x m)
#    p : initial state probabilities (length m)
#    B : observation probability matrix (m x k)
# Output:
#    ll : likelihood of the observation sequence
Dis.forward <- function(x, A, B, p) {
  m <- nrow(A)
  n <- length(x)
  # Forward recursion
  phi <- as.double(p)
```

```
  for (i in 1:n) {
    if (i > 1)
      phi <- phi %*% A
    phi <- phi * B[, x[i]]
  }
  # Final likelihood
  ll <- sum(phi)
  return(ll)
}
```

Now we will use the Dis.forward function to calculate the likelihood of observations in the HMM model from Example 1.2 in Section 1.3.4. First, we transform the observation sequence to a vector of binary value:

$$O = \{O_1, O_2, O_3\} = \{\text{Red, Blue, Blue}\} = \{\nu_1, \nu_2, \nu_2\} = \{1, 2, 2\}.$$

```
# Load the discrete forward algorithm function
source("C:/Users/Dis.forward.R")
# Define observation sequence and convert to numeric states
observations <- c("Red", "Blue", "Blue")
x <- ifelse(observations == "Red", 1, 2)
# Define transition and emission matrices
A <- matrix(c(0.7, 0.2,
              0.3, 0.8), ncol = 2, byrow = TRUE)
B <- matrix(c(0.75, 0.4,
              0.25, 0.6), ncol = 2, byrow = TRUE)
# Initial state probabilities
p <- c(1, 0)
# Compute likelihood using discrete forward algorithm
P_O_lambda <- Dis.forward(x, A, B, p)
cat("P(O|lambda)=", P_O_lambda, "\n")
```

The output is:

```
P(O | lambda) = 0.1181437.
```

The result is exactly what we got when calculated by hand. Note that in the code, we don't need to store the forward probability in every time step. We only need the forward probability at the last time period $t = T$ to calculate the likelihood $\mathbb{P}(O|\lambda)$. Therefore, we don't store the whole forward vector.

For a continuous HMM, recall that the emission probability of an observation is formulated from a continuous distribution, normally a Gaussian distribution. Therefore, we just need to modify the codes for the discrete HMM by calculating the emission probability based on the chosen emission distribution function. In the following code, we assume that the emission distributions are Gaussian distributions:

```
# Cont.forward function calculates the likelihood of a continuous
# HMM using the forward algorithm.
# Input:
#   x      : observation sequence (length n)
#   A      : transition matrix (m x m)
#   p      : initial state probabilities (length m)
```

```
#    mu     : vector of Gaussian means (length m)
#    sigma  : vector of Gaussian standard deviations (length m)
# Output:
#    ll     : likelihood of the observation sequence
Cont.forward <- function(x, A, p, mu, sigma) {
  m <- nrow(A)
  n <- length(x)
  # Compute emission probabilities
  prob <- matrix(0, nrow = n, ncol = m)
  for (k in 1:m)
    prob[, k] <- dnorm(x, mean = mu[k], sd = sigma[k], log = FALSE)
  # Forward recursion
  phi <- as.double(p)
  for (i in 1:n) {
    if (i > 1)
      phi <- phi %*% A
    phi <- phi * prob[i, ]
  }
  # Final likelihood
  ll <- sum(phi)
  return(ll)
}
```

Applying the Cont.forward function to calculate the likelihood of observation from Example 1.6 from Chapter 1:

```
source("C:/Users/Cont.forward.R") # Call the function to use
A <- matrix(c(0.4, 0.3, 0.3,
              0.2, 0.5, 0.3,
              0.6, 0.2, 0.2), ncol = 3, byrow = TRUE)
mu <- c(60, 55, 50)
sigma <- c(3, 2, 2.5)
p <- c(1, 0, 0)
x <- c(61, 57, 49)
L <- Cont.forward(x, A, p, mu, sigma)
sprintf("%.6f is the likelihood.", L)
```

The generated result: "0.000393 is the likelihood", which is the same result that we found in the solution to Example 1.6.

2.3 Backward Algorithm

Similar to these forward algorithms, we can use backward algorithms to calculate the probability $\mathbb{P}(O|\lambda)$. We will present these backward algorithms for single and multiple observation sequences, examples and R code to demonstrate these algorithms in this section.

2.3.1 Backward Algorithm for a Single Observation Sequence

Let's denote the backward probability function at the latest time T as

$$\beta_T(i) = 1, i = 1, ..., N, \tag{2.4}$$

and let backward probability functions at time $t < T$ be defined as

$$\beta_t(i) = P(O_{t+1}, O_{t+2}, ..., O_T | q_t = S_i, \lambda) \tag{2.5}$$

for $t = 1, ..., T-1$ and $i = 1, ..., N$. Then, we can calculate $\beta_t(i)$ by using the backward recursive formula:

$$\beta_t(i) = \sum_{j=1}^{N} a_{ij} b_j(O_{t+1}) \beta_{t+1}(j), t = T-1, T-2, ..., 1, i = 1, ..., N. \tag{2.6}$$

The conditional probability $\beta_t(i)$ is displayed in Figure 2.2.

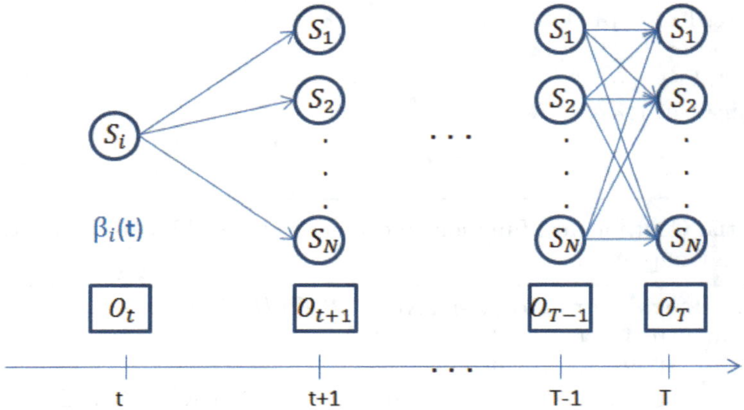

FIGURE 2.2
Backward probability function.

Algorithm 3 is the backward algorithm for one observation sequence.

Algorithm 3 The backward algorithm of the HMM with single observation sequence.

1: Initialization: for $i = 1, ..., N$

$$\beta_T(i) = 1.$$

2: Recursion: for $t = T-1, T-2, ..., 1$, for $i = 1, ..., N$

$$\beta_t(i) = \sum_{j=1}^{N} a_{ij} b_j(O_{t+1}) \beta_{t+1}(j).$$

3: Output:

$$\mathbb{P}(O|\lambda) = \sum_{\mathbf{i=1}}^{\mathbf{N}} \mathbf{p_i b_i(O_1)} \beta_1(\mathbf{i}).$$

Similar to the forward algorithm, using the backward algorithm we only need to do N^2T calculations to get the probability of observation $\mathbb{P}(O|\lambda)$.

2.3.2 Backward Algorithm for Multiple Observation Sequences

Now we will present the backward algorithm for multiple observation sequences $O = \{O^{(1)}, O^{(2)}, ..., O^{(L)}\}$. Let's denote the backward probability function at the latest time T as

$$\beta_T^{(l)}(i) = 1, i = 1, ..., N, l = 1, ..., L \tag{2.7}$$

and let backward probability functions at time $t < T$ be defined as

$$\beta_t^{(l)}(i) = P(O_{t+1}^{(l)}, O_{t+2}^{(l)}, ..., O_T^{(l)}|q_t = S_i, \lambda), \tag{2.8}$$

for $t = 1, ..., T-1$, $i = 1, ..., N$, and $l = 1, ..., L$.
Then, we can calculate $\beta_t^{(l)}(i)$ by using the backward recursive formula:

$$\beta_t^{(l)}(i) = \sum_{j=1}^{N} a_{ij} b_j(O_{t+1}^{(l)}) \beta_{t+1}^{(l)}(j), t = T-1, T-2, ..., 1, i = 1, ..., N, l = 1, ..., L. \tag{2.9}$$

Algorithm 4 is the backward algorithm for multiple independent observation sequences.

Algorithm 4 The backward algorithm for independent multiple observation sequences.

Initialization $\mathbb{P}(O|\lambda) = \mathbf{1}$

For $l = 1, 2, ..., L$ do

Initialization: for i=1,2,..., N

$$\beta_T^{(l)}(i) = 1.$$

Recursion: for $t = T-1, T-2, \ldots, T$, and for $j = 1, 2, \ldots, N$, compute

$$\beta_t^{(l)}(i) = \sum_{j=1}^{N} a_{ij} b_j(O_{t+1}^{(l)}) \beta_{t+1}^{(l)}(j).$$

Calculate:

$$\mathbb{P}(O^{(l)}|\lambda) = \sum_{\mathbf{i=1}}^{\mathbf{N}} \beta_1^{(\mathbf{l})}(\mathbf{i}).$$

Update:

$$\mathbb{P}(O|\lambda) = \mathbb{P}(\mathbf{O}|\lambda) \times \mathbb{P}(\mathbf{O}^{(\mathbf{l})}|\lambda).$$

Output: $\mathbb{P}(O|\lambda)$.

2.3.3 Example of Using the Backward Algorithm

We now apply the backward algorithm to calculate the probability of observations, $\mathbb{P}(O|\lambda)$, in the "Ball Game" example from Section 1.3.4. Recall that the observation sequence is

$$O = \{O_1, O_2, O_3\} = \{\text{Red, Blue, Blue}\}.$$

The transition probability matrix, the observation probability matrix, and the initial probability of being in each state are respectively

$$\mathbf{A} = \begin{bmatrix} 0.7 & 0.3 \\ 0.2 & 0.8 \end{bmatrix}, \mathbf{B} = \begin{bmatrix} 0.75 & 0.25 \\ 0.40 & 0.60 \end{bmatrix}, \textbf{ and } \mathbf{p} = (\mathbf{p_1}, \mathbf{p_2}) = (\mathbf{1}, \mathbf{0}).$$

The model parameter is $\lambda = \{\mathbf{A}, \mathbf{B}, \mathbf{p}\}$, number of states $N = 2, S_1 = \text{Red}, S_2 = \text{Blue}$, and number of observations is $T = 3$. First, we will calculate the backward function at time t . By the backward algorithm 3, at time $T = 3$, we have:

$$\beta_3(1) = \beta_3(2) = 1.$$

At time $t = 2$, we have:

$$\begin{aligned} \beta_2(1) &= \beta_3(1)a_{11}b_1(O_3) + \beta_3(2)a_{12}b_2(O_3) \\ &= 1(0.7)(0.25) + 1(0.3)(0.6) \\ &= 0.335 \\ \beta_2(2) &= \beta_3(1)a_{21}b_1(O3) + \beta_3(2)a_{22}b_2(O_3) \\ &= 1(0.2)(0.25) + 1(0.8)(0.6) \\ &= 0.53. \end{aligned}$$

At time $t = 1$, we have:

$$\begin{aligned} \beta_1(1) &= \beta_2(1)a_{11}b_1(O_2) + \beta_2(2)a_{12}b_2(O_2) \\ &= (0.335)(0.7)(0.25) + 0.53(0.3)(0.8) \\ &= 0.157525 \\ \beta_1(2) &= \beta_2(1)a_{21}b_1(O_2) + \beta_2(2)a_{22}b_2(O_2) \\ &= (0.335)(0.2)(0.25) + 0.53(0.8)(0.6) \\ &= 0.27215. \end{aligned}$$

Finally, the observation probability is calculated as

$$\begin{aligned} \mathbb{P}(O|\lambda) &= p_1b_1(O_1)\beta_1(1) + p_2b_2(O_1)\beta_1(2) \\ &= 1(0.75)(0.157525) + 0 \\ &= 0.118144. \end{aligned}$$

The result is identical with the result using the forward algorithm. We also notice that

$$\mathbb{P}(O, q_t = S_i|\lambda) = \alpha_\mathbf{t}(\mathbf{i})\beta_\mathbf{t}(\mathbf{i}). \tag{2.10}$$

Thus, we can use the forward and backward functions to calculate the model's likelihood:

$$\mathbb{P}(O|\lambda) = \sum_{\mathbf{i=1}}^{\mathbf{N}} \mathbb{P}(\mathbf{O}, \mathbf{q_t} = \mathbf{S_i}|\lambda) = \sum_{\mathbf{i=1}}^{\mathbf{N}} \alpha_\mathbf{t}(\mathbf{i})\beta_\mathbf{t}(\mathbf{i}), \tag{2.11}$$

for any $1 \le t \le T$. Therefore, we can apply the equation (2.11) for $t = 1, 2, \text{or } 3$ to calculate $\mathbb{P}(O|\lambda)$. For example, with $t = 2$, we have

$$\begin{aligned} \mathbb{P}(O|\lambda) &= \alpha_2(1)\beta_2(1) + \alpha_2(2)\beta_2(2) \\ &= 0.13125(0.355) + 0.135(0.53) \\ &= 0.118144. \end{aligned}$$

The properties of the forward and backward probability functions will be used to calculate the probability of observation in the Baum-Welch algorithms. We will present the Baum-Welch algorithm in Section 2.5.

2.3.4 R Code for the Backward Algorithm

We present the R code to calculate the probability of observations using the backward algorithm for a discrete HMM:

```
# Dis.backward function calculates the likelihood of a discrete
# HMM using the backward algorithm.
# Input:
#     x : observation sequence (length n)
#     A : transition matrix (m x m)
#     p : initial state probabilities (length m)
#     B : observation probability matrix (m x k)
# Output:
#     ll : likelihood of the observation sequence
Dis.backward <- function(x, A, B, p) {
  m <- nrow(A)
  n <- length(x)
  # Initialize backward matrix
  beta <- matrix(0, nrow = n, ncol = m)
  beta[n, ] <- rep(1, m)
  # Backward recursion
  for (i in seq(n - 1, 1, -1))
    beta[i, ] <- A %*% (B[, x[i + 1]] * beta[i + 1, ])
  # Final likelihood
  ll <- sum(p * B[, x[1]] * beta[1, ])
  return(ll)
}
```

Using the Dis.backward code to calculate the likelihood of observations from Example 1.2:

```
# Call the function to use
source("C:/Users/Dis.backward.R")
# Define transition and emission matrices
A <- matrix(c(0.7, 0.2,
              0.3, 0.8), ncol = 2, byrow = TRUE)
B <- matrix(c(0.75, 0.4,
              0.25, 0.6), ncol = 2, byrow = TRUE)
# Initial state probabilities and observation sequence
p <- c(1, 0)
x <- c(1, 2, 2)
# Compute likelihood using discrete backward algorithm
L <- Dis.backward(x, A, B, p)
sprintf("%f is the Likelihood", L)
```

we got the result: "0.118144 is the Likelihood." Similar to the forward codes, the backward codes gave us the same the likelihood $P(O|\lambda) = 0.118144$.

For continuous HMM, we have the following code for the backward algorithm:

```
# Cont.backward function calculates the likelihood of a continuous
# HMM using the backward algorithm.
# Input:
#     x      : observation sequence (length n)
```

```
#   A     : transition matrix (m x m)
#   p     : initial state probabilities (length m)
#   mu    : vector of Gaussian means (length m)
#   sigma : vector of Gaussian standard deviations (length m)
# Output:
#   ll    : likelihood of the observation sequence
Cont.backward <- function(x, A, p, mu, sigma) {
  m <- nrow(A)
  n <- length(x)
  # Compute emission probabilities
  prob <- matrix(0, nrow = n, ncol = m)
  for (k in 1:m)
    prob[, k] <- dnorm(x, mean = mu[k], sd = sigma[k], log = FALSE)
  # Backward recursion
  beta <- matrix(0, nrow = n, ncol = m)
  beta[n, ] <- rep(1, m)
  for (i in seq(n - 1, 1, -1))
    beta[i, ] <- A %*% (prob[i + 1, ] * beta[i + 1, ])
  # Final likelihood
  ll <- sum(p * prob[1, ] * beta[1, ])
  return(ll)
}
```

Using the Cont.backward function to calculate the probability of observations from Example 1.6:

```
source("C:/Users/Cont.backward.R") # Call the function to use
A <- matrix(c(0.4, 0.3, 0.3,
              0.2, 0.5, 0.3,
              0.6, 0.2, 0.2), ncol = 3, byrow = TRUE)
mu <- c(60, 55, 50)
sigma <- c(3, 2, 2.5)
p <- c(1, 0, 0)
x <- c(61, 57, 49)
L <- Cont.backward(x, A, p, mu, sigma)
sprintf("%#.6f is the likelihood", L)
```

we got the identical likelihood that we found from Example 1.6.

2.4 Viterbi Algorithm

The Viterbi algorithm is used to solve the second problem of the HMM: find the "best fit" hidden states of the observations. The goal here is to find the best sequence of states Q when the model parameters and observations, (O, λ), are given to maximize the probability of observation $\mathbb{P}(O|\lambda)$. There are many possible state sequences Q for an observation sequence O. Among these state sequences, we need to find the "best fit" sequence Q^* that satisfies the maximization probability

$$\mathbb{P}(O, Q^*|\lambda) = \max_{\mathbf{allQ}} \mathbb{P}(\mathbf{O}, \mathbf{Q}|\lambda). \tag{2.12}$$

2.4.1 Viterbi Algorithm for a Single Observation Sequence

The process of selecting the hidden states is explained step by step in the Viterbi algorithm for a single observation sequence. We define

$$\delta_t(i) = \max_{q_1,q_2,\ldots,q_{t-1}} \mathbb{P}(q_1, q_2, \ldots, q_t = S_i, O_1, \ldots, O_t|\lambda). \tag{2.13}$$

Then, the maximum of observation probability with an optimal state sequence Q^* is the maximum of function $\delta_T(i)$ on all possible values j,

$$\mathbb{P}(O, Q^*|\lambda) = \max_{\mathbf{1 \le j \le N}} \delta_{\mathbf{T}}(\mathbf{i}). \tag{2.14}$$

The function $\delta_t(i)$ can be calculated recursively by initializing its values at time $t = 1$ as

$$\delta_1(j) = p_j b_j(O_1), 1 \le j \le N$$

and using the recursive formula

$$\delta_{t+1}(j) = b_j(O_{t+1}) \max_i \{\delta_t(i)a_{ij}\}. \tag{2.15}$$

Using $\delta_t(i)$, we can locate the most likely state q_t, at time t, as

$$q_t = \arg \max_{1 \le i \le N} [\delta_t(i)], \quad 1 \le t \le T. \tag{2.16}$$

Thus, the Viterbi algorithm for a single observation sequence is given below.

The Viterbi algorithm for a single observation sequence

1: Initialization:

$$\delta_1(j) = p_j b_j(O_1), \quad j = 1, 2, ..., N;$$

$$\phi_1(j) = 0.$$

2: Recursion: for $2 \le t \le T$, and $1 \le j \le N$

$$\begin{aligned} \delta_t(j) &= b_j(O_{t+1}) \max_i \{\delta_{t-1}(i)a_{ij}\} \\ \phi_t(j) &= \arg\max_i [\delta_{t-1}(i)a_{ij}]. \end{aligned}$$

3: Probability of observation in the hidden state

$$P = \max_{1 \le j \le N} \delta_T(i).$$

4: State prediction

$$\begin{aligned} q_T^* &= argmax_i[\delta_T(i)] \\ q_t^* &= \phi_{t+1}(q_{t+1}^*), \; t = T-1, \ldots, 1. \end{aligned}$$

2.4.2 Example of Using the Viterbi Algorithm

We can now apply the Viterbi algorithm functions in Section 2.4.1 to find the "best fit" hidden states Q^* (or the sequence of boxes) of the observations $O = \{\text{Red, Blue, Blue}\}$ in Example 1.4 from Section 1.3.4 to maximize the probability of observations $P(O, Q^*|\lambda$.

Recall that the transition probability matrix, the observation probability matrix, and the initial probability of being in each state are respectively

$$\mathbf{A} = \begin{bmatrix} 0.7 & 0.3 \\ 0.2 & 0.8 \end{bmatrix}, \mathbf{B} = \begin{bmatrix} 0.75 & 0.25 \\ 0.40 & 0.60 \end{bmatrix}, \text{ and } \mathbf{p} = (\mathbf{1}, \mathbf{0}).$$

To find the optimal hidden sequence Q^* using the Viterbi algorithm, we will calculate values of the probability function $\delta_t(j)$ and the index function $\Phi_t(j)$ at each time t recursively. At time $t = 1$, we have:

$$\begin{aligned} \delta_1(1) &= p_1 b_1(O_1) = p_1 b_1(Red) = p_1 b_{11} = 1(0.75) = .75 \\ \delta_1(2) &= p_2 b_2(O_1) = p_2 b_2(Red) = p_2 b_{21} = 0(0.4) = 0 \\ \Phi_1(1) &= \Phi_1(2) = 0. \end{aligned}$$

At time $t = 2$, we have:

$$\begin{aligned} \delta_2(1) &= b_1(O_2) \max_{i=1,2} \delta_1(i) a_{i1} = b_1(O_2)\delta_1(1)a_{11} \\ &= (0.25)0.75(0.7) = 0.13125 \\ \delta_2(2) &= b_2(O_2) \max_{i=1,2}\{\delta_1(i)a_{i2}\} = b_2(O_2)\delta_1(1)a_{12} \\ &= .25(0.75)(0.3) = .05625 \\ \Phi_2(1) &= \arg\max_{i=1,2}\{\delta_1(i)a_{i1}\} = \arg\max_{i=1,2}\{.71(.7), 0(.2)\} = 1 \\ \Phi_2(2) &= \arg\max_{i=1,2}\{\delta_1(i)a_{i2}\} = \arg\max_{i=1,2}\{.71(.3), 0(.8)\} = 1. \end{aligned}$$

At time $t = 3$, we have:

$$\begin{aligned} \delta_3(1) &= b_1(O_3) \max_{i=1,2}\{\delta_2(i)a_{i1}\} = b_1(O_3)\delta_2(1)a_{11} \\ &= 0.13125(0.7)(0.25) = 0.02297 \\ \delta_3(2) &= b_2(O_3) \max_{i=1,2}\{\delta_2(i)a_{i2}\} = b_2(O_3)\delta_2(1)a_{12} \\ &= 0.13125(0.3)(0.6) = 0.02363 \\ \Phi_3(1) &= \arg\max_{i=1,2}\{\delta_2(i)a_{i1}\} = \arg\max_{i=1,2}\{0.13125(.7), .05625(.2)\} = 1 \\ \Phi_3(2) &= \arg\max_{i=1,2}\{\delta_2(i)a_{i2}\} = \arg\max_{i=1,2}\{0.13125(.3), .05625(.8)\} = 2. \end{aligned}$$

Thus, the "best fit" hidden states are:

$$\begin{aligned} q_3^* &= \arg\max_{i=1,2}\{\delta_3(i)\} = \arg\max_{i=1,2}\{0.02297, 0.02363\} = 2 \\ q_2^* &= \Phi_3(q_3^*) = \Phi_3(2) = 2 \\ q_1^* &= \Phi_2(q_2^*) = \Phi_2(2) = 1. \end{aligned}$$

In summary, the "best fit" state sequence is

$$Q^* = \{q_1^*, q_2^*, q_3^*\} = \{1, 2, 2\} = \{S_1, S_2, S_2\} = \{\text{Box 1,Box 2,Box 2}\}.$$

This result agrees with the result that we found in Example 1.4 in Section 1.3.4.

2.4.3 R Code for the Viterbi Algorithm

Now we can check our calculations by running the R code of the Viterbi algorithm:

```
# Dis.Viterbi function finds the most likely hidden state
# sequence for a discrete HMM using the Viterbi algorithm.
# Input:
#   x : observation sequence (length n)
#   A : transition matrix (m x m)
#   p : initial state probabilities (length m)
#   B : observation probability matrix (m x k)
# Output:
#   Q : most likely hidden state sequence (length n)
Dis.Viterbi <- function(A, B, p, x) {
  m <- nrow(A)
  T <- length(x)
  # Initialize Viterbi matrices
  Q <- integer(T)
  theta <- matrix(0, nrow = T, ncol = m)
  Phi <- matrix(0, nrow = T, ncol = m)
  # Initialization
  theta[1, ] <- p * B[, x[1]]
  # Recursion
  for (t in 2:T) {
    for (i in 1:m) {
      scores <- theta[t - 1, ] * A[, i]
      theta[t, i] <- max(scores) * B[i, x[t]]
      Phi[t, i] <- which.max(scores)
    }
  }
  # Backtracking
  Q[T] <- which.max(theta[T, ])
  for (t in (T - 1):1)
    Q[t] <- Phi[t + 1, Q[t + 1]]
  return(Q)
}
```

Using the Dis.Viterbi function to solve the Example 1.2 in Chapter 1:

```
# Load the discrete Viterbi algorithm function
source("C:/Users/Dis.Viterbi.R")
# Define transition and emission matrices
A <- matrix(c(0.7, 0.2,
              0.3, 0.8), ncol = 2, byrow = TRUE)
B <- matrix(c(0.75, 0.4,
              0.25, 0.6), ncol = 2, byrow = TRUE)
# Initial state probabilities and observation sequence
p <- c(1, 0)
x <- c(1, 2, 2)
# Compute most likely hidden state sequence
S <- Dis.Viterbi(A, B, p, x)
print("Hidden state sequence is Q =")
print(S)
```

we got the R output: "Hidden state sequence is Q = 1 2 2". The output result is the state sequence $Q = \{1, 2, 2\}$ = {Box 1, Box 2, Box 2}. You can try to run the codes with different input to see the results.

For continuous HMM, we have the following Cont.Viterbi function of using the Viterbi algorithm to find the optimal hidden states:

```
# Cont.Viterbi function finds the most likely hidden state
# sequence for a continuous HMM using the Viterbi algorithm.
# Input:
#   x     : observation sequence (length n)
#   A     : transition matrix (m x m)
#   p     : initial state probabilities (length m)
#   mu    : vector of Gaussian means (length m)
#   sigma : vector of Gaussian standard deviations (length m)
# Output:
#   Q     : most likely hidden state sequence
Cont.Viterbi <- function(x, A, p, mu, sigma) {
  m <- nrow(A)
  n <- length(x)
  # Compute emission probabilities
  prob <- matrix(0, nrow = n, ncol = m)
  for (k in 1:m)
    prob[, k] <- dnorm(x, mean = mu[k], sd = sigma[k], log = FALSE)
  # Initialize Viterbi matrices
  Q <- integer(n)
  theta <- matrix(0, nrow = n, ncol = m)
  Phi <- matrix(0, nrow = n, ncol = m)
  theta[1, ] <- p * prob[1, ]
  # Recursion
  for (t in 2:n) {
    for (i in 1:m) {
      scores <- theta[t - 1, ] * A[, i]
      theta[t, i] <- max(scores) * prob[t, i]
      Phi[t, i] <- which.max(scores)
    }
  }
  # Backtracking
  Q[n] <- which.max(theta[n, ])
  for (t in (n - 1):1)
    Q[t] <- Phi[t + 1, Q[t + 1]]
  return(Q)
}
```

Using the Cont.Viterbi function to solve Example 1.7 from Chapter 1:

```
# Load the continuous Viterbi algorithm function
source("C:/Users/Cont.Viterbi.R")
# Define transition matrix
A <- matrix(c(0.4, 0.3, 0.3,
              0.2, 0.5, 0.3,
              0.6, 0.2, 0.2), ncol = 3, byrow = TRUE)
# Define Gaussian emission parameters
```

```
mu <- c(60, 55, 50)
sigma <- c(3, 2, 2.5)
# Initial state probabilities and observation sequence
p <- c(1, 0, 0)
x <- c(61, 57, 49)
# Compute most likely hidden state sequence
Q <- Cont.Viterbi(x, A, p, mu, sigma)
print(Q)
```

we achieved the output: $Q = \{1, 2, 3\}$, which is equivalent to $Q = Q_6 = \{S_1, S_2, S_3\} = \{\text{Sunny, Rainy, Cloudy}\}$. The result is identical with the solution of the Example 1.7 in Chapter 1.

2.5 Baum-Welch Algorithm

Finally, we turn to the third problem of the HMM presented in Section 2.1, which is the most difficult problem of the three problems. In the third problem, we have to find the parameters $\lambda = \{\mathbf{A}, \mathbf{B}, \mathbf{p}\}$ given an observation sequence O. A fundamental method is finding a set of parameters λ that maximizes the probability of observation $\mathbb{P}(O|\lambda)$. Unfortunately, given observation data, there is no way to find a global maximum of $\mathbb{P}(O|\lambda)$. However, we can find a local maximum of $\mathbb{P}(O|\lambda)$ using the Baum-Welch algorithm [13]. The Baum-Welch algorithm is an algorithm to calibrate parameters for the HMM given the observation data. The process of the Baum-Welch algorithm is as follows: with an initial set of parameters λ, the algorithm estimates a new parameter λ^* using an optimization method to maximize the probability of observation $\mathbb{P}(O|\lambda)$ where $\mathbb{P}(O|\lambda^*) > \mathbb{P}(\mathbf{O}|\lambda)$. We will terminate the process when we reach a desired error or number of interactions. The re-estimate parameters' process of the HMM with single observation sequence and multiple observation sequences are explained in the next section.

2.5.1 Baum-Welch Algorithm for a Single Observation Sequence

The Baum-Welch method, or the expectation modification (EM) method, is used to find a local maximizer, λ^*, of the probability function $\mathbb{P}(O|\lambda)$. By definitions of the forward and backward probability functions, the observation probability $\mathbb{P}(O|\lambda)$ can be calculated by using both forward and backward functions, $\alpha_t(i)$ and $\beta_t(i)$, at any time t,

$$\mathbb{P}(O|\lambda) = \sum_{1}^{\mathbf{N}} \mathbb{P}(\mathbf{q_t} = \mathbf{S_i}|\mathbf{O}, \lambda) = \sum_{1}^{\mathbf{N}} \alpha_\mathbf{t}(\mathbf{i})\beta_\mathbf{t}(\mathbf{i}). \tag{2.17}$$

In the re-estimate parameters' procedure, we define the function $\gamma_t(i)$, the probability of being in state S_i at time t, as:

$$\gamma_t(i) = \mathbb{P}(q_t = S_i|O, \lambda), \tag{2.18}$$

for $i = 1, ..., N$.
By applying Bayes' rule, we have:

$$\gamma_t(i) = \frac{P(q_t = S_i, O|\lambda)}{\mathbb{P}(O|\lambda)} = \frac{\alpha_t(i)\beta_t(i)}{\mathbb{P}(O|\lambda)} = \frac{\alpha_t(i)\beta_t(i)}{\sum_{i=1}^{N} \alpha_t(i)\beta_t(i)}. \tag{2.19}$$

The probability of being in state S_i at time t and state S_j at time $t+1$, $\xi_t(i,j)$ is defined as:

$$\xi_t(i,j) = \mathbb{P}(q_t = S_i, q_{t+1} = S_j | O, \lambda). \tag{2.20}$$

Similarly, using Bayes' rule, we have:

$$\xi_t(i,j) = \frac{\mathbb{P}(q_t = S_i, q_{t+1} = S_j, O|\lambda)}{\mathbb{P}(O|\lambda)}. \tag{2.21}$$

The numerator can be split to the two terms

$$\mathbb{P}(q_t = S_i, q_{t+1} = S_j, O|\lambda) = \mathbb{P}(\mathbf{O_1, ..., O_t, q_t = S_i}|\lambda) \cdot \mathbb{P}(\mathbf{O_{t+1}, ..., O_T, q_{t+1} = S_j | q_t = S_i}, \lambda).$$

By the definition of the forward function in (2.1), we have:

$$\mathbb{P}(q_t = S_i, q_{t+1} = S_j, O|\lambda) = \alpha_\mathbf{t}(\mathbf{i}) \cdot \mathbb{P}(\mathbf{O_{t+1}, ..., O_T, q_{t+1} = S_j | q_t = S_i}, \lambda).$$

Furthermore, the conditional probability can be calculated as:

$$\mathbb{P}(O_{t+1}, ..., O_T, q_{t+1} = S_j | q_t = S_i, \lambda) = \mathbb{P}(\mathbf{O_{t+1}, q_{t+1} = S_j | q_t = S_i}, \lambda) \cdot \mathbb{P}(\mathbf{O_{t+1}, ..., O_T, q_{t+1} = S_j}|\lambda).$$

By the definition of the backward function in (2.5), we have:

$$\mathbb{P}(q_t = S_i, q_{t+1} = S_j, O|\lambda) = \alpha_\mathbf{t}(\mathbf{i})\mathbf{a_{ij} b_j (O_{t+1})}\beta_\mathbf{t+1}(\mathbf{j}).$$

Thus, the probability function $\xi_t(i,j)$ can be calculated by

$$\xi_t(i,j) = \frac{\alpha_t(i) a_{ij} b_j(O_{t+1}) \beta_{t+1}(j)}{\mathbb{P}(O|\lambda)}. \tag{2.22}$$

The break out formula to calculate the probability function $\xi_t(i,j)$ in equation (2.22) is displayed in Figure 2.3.

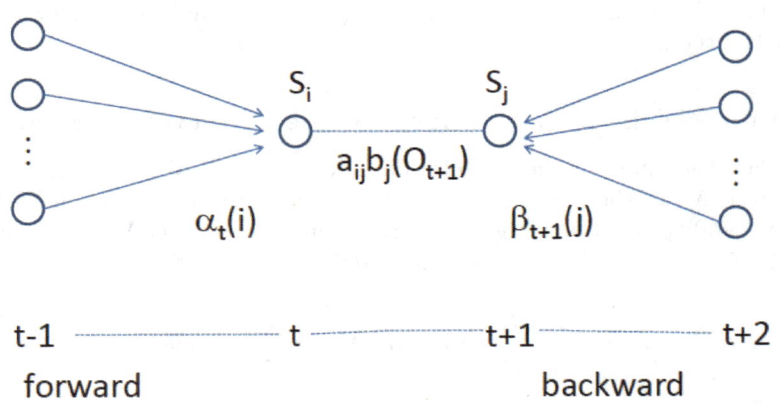

FIGURE 2.3
Components of $\xi_t(i,j)$ function.

By definitions of $\gamma_t(i)$ in (2.18) and the function $\xi_t(i,j)$ in (2.20), we have

$$\gamma_t(i) = \sum_{j=1}^{N} \xi_t(i,j). \tag{2.23}$$

We can consider that $\gamma_t(i)$ is the expected number of times that state s_i is visited, then, we have that $\sum_{t=1}^{T-1} \gamma_t(i)$ is the expected number of transitions made from state S_i, and then $\sum_{t=1}^{T-1} \xi_t(i,j)$ is the expected number of transitions from state S_i to S_j. We can use the following formulas to update parameter λ^* of a discrete HMM:

$$p_i^* = \gamma_1(i), 1 \leq i \leq N \tag{2.24}$$

$$a_{ij}^* = \frac{\sum_{t=1}^{T-1} \xi_t(i,j)}{\sum_{t=1}^{T-1} \gamma_t(i)} \tag{2.25}$$

$$b_{jk}^* = b_j^*(k) = \frac{\sum_{t=1,|O_t=\nu_k}^{T} \gamma_t(j)}{\sum_{t=1}^{T} \gamma_t(i)}. \tag{2.26}$$

The constraints of the model parameters $\lambda^* = \{\mathbf{A}^*, \mathbf{B}^*, \mathbf{p}^*\}$:

$$a_{ij}^* > 0, \sum_{j=1}^{N} a_{ij}^* = 1, 1 \leq i,j \leq N. \tag{2.27}$$

$$b_{ik}^* > 0, \sum_{k=1}^{M} b_i^*(k) = 1, 1 \leq j \leq N, 1 \leq k \leq M. \tag{2.28}$$

$$p_i^* > 0, \sum_{i=1}^{N} p_i^* = 1, 1 \leq i \leq N. \tag{2.29}$$

are automatically satisfied because of the definition of $\gamma_t(i)$ in (2.18) and its property in (2.23).

The authors in [13] proved that with an initial parameter λ^*, each time we update the model's parameters λ^* using the aforementioned formulas, the probability of observation increases, $\mathbb{P}(O^*|\lambda) > \mathbb{P}(\mathbf{O}|\lambda)$. The complete algorithm 5 is the Baum-Welch algorithm for a discrete HMM with one sequence of observations.

Algorithm 5 Baum-Welch algorithm for a discrete HMM with one observation sequence.

1: Initialization: input parameters λ, the tolerance *tol*, and a real number $\triangle$
2: Repeat until $\triangle < tol$
3: Calculate $\mathbb{P}(O|\lambda)$ using the forward algorithm in Section 2.1
4: Calculate new parameters λ^*: for $1 \leq i \leq N$

$$p_i^* = \gamma_1(i)$$

$$a_{ij}^* = \frac{\sum_{t=1}^{T-1} \xi_t(i,j)}{\sum_{t=1}^{T-1} \gamma_t(i)}, \quad 1 \leq j \leq N$$

$$b_{ik}^* = \frac{\sum_{t=1}^{T} I_{O_t=v_k} \gamma_t(i)}{\sum_{t=1}^{T} \gamma_t(i)}, \quad 1 \leq k \leq M.$$

5: Calculate $\triangle = |\mathbb{P}(O|\lambda^*) - \mathbb{P}(\mathbf{O}|\lambda)|$.
6: Update $\lambda = \lambda^*$.
7: Output: parameters λ.

For a continuous HMM, we present the calculations of the new parameter vector, λ^*, for the case when the probability of observation at time t, b_{ik}^*, is a $M-$ mixture Gaussian in equation (1.23) as mentioned in Section 1.3.2,

$$b_i(O_t) = \sum_{m=1}^{M} c_{im}\mathcal{N}(O_t, \mu_{im}, \sigma_{im}), 1 \leq i \leq N.$$

The re-estimate parameters $c_{ik}, \mu_{ik}, \sigma_{ik}$ for the mixture Gaussian that was presented in ([90], [71], [72]) are:

$$c^*_{ik} = \frac{\sum_{t=1}^{T} \gamma_t(i,k)}{\sum_{t=1}^{T}\sum_{k=1}^{M} \gamma_t(i,k)} \tag{2.30}$$

$$\mu^*_{ik} = \frac{\sum_{t=1}^{T} \gamma_t(i,k)O_t}{\sum_{t=1}^{T} \gamma_t(i,k)} \tag{2.31}$$

$$\sigma^*_{ik} = \frac{\sum_{t=1}^{T} \gamma_t(i,k)(O_t - \mu_i)(O_t - \mu_{ik})'}{\sum_{t=1}^{T} \gamma_t(i,k)}, \tag{2.32}$$

where $(O_t - \mu_{ik})'$ is the transpose of $(O_t - \mu_{ik})$, $1 \leq i \leq N$, $1 \leq k \leq M$, and $\gamma_t(i,k)$ is the probability of being in state i at time t with k^{th} mixture component; accounting for O_t defined by

$$\gamma_t(i,k) = \left[\frac{\alpha_t(i)\beta_t(i)}{\sum_{i=1}^{N} \alpha_t(i)\beta_t(i)}\right]\left[\frac{c_{ik}\mathcal{N}(O_t, \mu_{ik}, \sigma_{ik})}{\sum_{k=1}^{M} c_{ik}\mathcal{N}(O_t, \mu_{ik}, \sigma_{ik})}\right]. \tag{2.33}$$

When $M = 1$, we have the simple re-estimate formulas for μ and σ where $c = 1$,

$$\mu^*_i = \frac{\sum_{t=1}^{T} \gamma_t(i)O_t}{\sum_{t=1}^{T} \gamma_t(i)}$$
$$\sigma^*_i = \frac{\sum_{t=1}^{T} \gamma_t(i)(O_t - \mu_i)(O_t - \mu_i)'}{\sum_{t=1}^{T} \gamma_t(i)}, \tag{2.34}$$

where $(O_t - \mu_i)'$ is the transpose of $(O_t - \mu_i)$,and $1 \leq i \leq N$.

2.5.2 Baum-Welch Algorithm for Multiple Observation Sequences

In this section, we will present the Baum-Welch algorithm for L independent observations $O = (O^{(1)}, O^{(2)}, ..., O^{(L)})$ with the same length T. With minor changes to the definitions of the backward probability functions (2.5), we get the following definition of the backward functions

$$\beta^{(l)}_t(i) = \mathbb{P}(O^{(l)}_{t+1}, O^{(l)}_{t+2}, .., O^{(l)}_T | q_t = S_i, \lambda), \tag{2.35}$$

for $i = 1, ..., N, l = 1, 2, ..., L$.
Obviously, for $i = 1, 2, ..., N$, we have

$$\beta^{(l)}_T(i) = 1. \tag{2.36}$$

Thus, we have the following backward recursive:

$$\beta^{(l)}_t(i) = \sum_{j=1}^{N} a_{ij} b_j(O^{(l)}_{t+1})\beta^{(l)}_{t+1}(j), t = T-1, T-2, ..., 1. \tag{2.37}$$

We then define $\gamma_t^{(l)}(i)$, the probability of being in state S_i at time t of the observation $O^{(l)}$ for $l = 1, 2, ..., L$, as:

$$\gamma_t^{(l)}(i) = P(q_t = S_i | O^{(l)}, \lambda) = \frac{\alpha_{\mathbf{t}}^{(\mathbf{l})}(\mathbf{i})\beta_{\mathbf{t}}^{(\mathbf{l})}(\mathbf{i})}{\mathbb{P}(\mathbf{O}^{(\mathbf{l})}|\lambda)} = \frac{\alpha_{\mathbf{t}}^{(\mathbf{l})}(\mathbf{i})\beta_{\mathbf{t}}^{(\mathbf{l})}(\mathbf{i})}{\sum_{\mathbf{i}=\mathbf{1}}^{\mathbf{N}} \alpha_{\mathbf{t}}^{(\mathbf{l})}(\mathbf{i})\beta_{\mathbf{t}}^{(\mathbf{l})}(\mathbf{i})}. \tag{2.38}$$

The probability of being in state S_i at time t and state S_j at time $t + 1$ of the observation $O^{(l)}, l = 1, 2, ..., L$ as:

$$\xi_t^{(l)}(i, j) = P(q_t = S_i, q_{t+1} = S_j | O^{(l)}, \lambda) = \frac{\alpha_{\mathbf{t}}^{(\mathbf{l})}(\mathbf{i})\mathbf{a_{ij}}\mathbf{b_j}(\mathbf{O}_{\mathbf{t+1}}^{(\mathbf{l})})\beta_{\mathbf{t+1}}^{(\mathbf{l})}(\mathbf{j})}{\mathbb{P}(\mathbf{O}^{(\mathbf{l})}, \lambda)}. \tag{2.39}$$

Clearly, we have

$$\gamma_t^{(l)}(i) = \sum_{j=1}^{N} \xi_t^{(l)}(i, j). \tag{2.40}$$

Algorithm 6 is the Baum-Welch algorithm for a discrete HMM with multiple independent observation sequences.

Algorithm 6 Baum-Welch algorithm for a discrete HMM with L independent observation sequences $\{O^{(1)}, O^{(2)}, ..., O^{(L)}\}$ with the same length T.

1: Initialization: input parameters λ, the tolerance δ, and a real number $\triangle$
2: Repeat until $\triangle < \delta$
3: Calculate $\mathbb{P}(O|\lambda) = \mathbf{\Pi}_{\mathbf{l=1}}^{\mathbf{L}}\mathbb{P}(\mathbf{O}^{(\mathbf{l})}|\lambda)$ using the Algorithm 1.
4: Calculate new parameters $\lambda^* = \{\mathbf{A}^*, \mathbf{B}^*, \mathbf{p}^*\}$, for $1 \le i \le N$

$$p_i^* = \frac{1}{L}\sum_{l=1}^{L} \gamma_1^{(l)}(i)$$

$$a_{ij}^* = \frac{\sum_{l=1}^{L}\sum_{t=1}^{T-1} \xi_t^{(l)}(i, j)}{\sum_{l=1}^{L}\sum_{t=1}^{T-1} \gamma_t^{(l)}(i)}, \quad 1 \le j \le N$$

$$b_i(k)^* = \frac{\sum_{l=1}^{L}\sum_{t=1}^{T} I_{O_t^{(l)}=v_k^{(l)}}\gamma_t^{(l)}(i)}{\sum_{l=1}^{L}\sum_{t=1}^{T} \gamma_t^{(l)}(i)}, \quad 1 \le k \le M$$

5: Calculate $\triangle = \mathbb{P}(O|\lambda^*) - \mathbb{P}(\mathbf{O}|\lambda)$
6: Update

$$\lambda = \lambda^*.$$

7: Output: parameters λ.

Note that the parameter λ^* was updated in Step 2 of the Baum-Welch algorithm to maximize the function $\mathbb{P}(O|\lambda)$ so that we have $\triangle = \mathbb{P}(O|\lambda^*) - \mathbb{P}(\mathbf{O}|\lambda) > \mathbf{0}$.

For a continuous HMM, we present the case that the observation probability $b_i(k)^*$, defined in equation (1.23) in Section 1.3.2, is $M-$ mixture Gaussian with $M = 1$. We will use the following formula to update the model parameter, $\lambda \equiv \{\mathbf{A}, \mu, \sigma, \mathbf{p}\}$

$$\mu_i^* = \frac{\sum_{l=1}^{L}\sum_{t=1}^{T}\gamma_t^{(l)}(i)O_t^{(l)}}{\sum_{l=1}^{L}\sum_{t=1}^{T-1}\gamma_t^{(l)}(i)},$$

$$\sigma_i^* = \frac{\sum_{l=1}^{L}\sum_{t=1}^{T}\gamma_t^{(l)}(i)(O_t^{(l)}-\mu_i)(O_t^{(l)}-\mu_i)'}{\sum_{l=1}^{L}\sum_{t=1}^{T}\gamma_t^{(l)}(i)}.$$

2.5.3 Example of Using the Baum-Welch Algorithm

In this example, we use the Baum-Welch algorithm to re-estimate the model's parameters presented in Example 1.5 from Section 1.3.4. First, we will restate the "Ball Game". Suppose that we have two boxes of balls, and each box contains two possible ball colors: red or blue. A player will first choose a box by using a process that satisfies the first order of a Markov chain with a constant transition probability matrix **A**. Then, the player will pick a ball from the selected box. Given the observations of three successive tries $O = \{\text{Red, Blue, Blue}\}$, can we find the matrix **A**, the probability of observing a red or blue ball in his first try (vector **p**), and the numbers of red and blue balls in each box (or the matrix **B**: probability of getting a red or blue ball from each box)? The answer is yes, we can use the Baum-Welch algorithm to find the parameters.

Recall that the discrete HMM has two states ($S_1 = \text{Box } 1, S_2 = \text{Box } 2$) and two symbols for each state are ($\nu_1 = \text{Red}, \nu_2 = \text{Blue}$). We choose the initial parameters for the model as the ones that we used in Example 1.2, $\lambda = \{\mathbf{A}, \mathbf{B}, \mathbf{p}\}$, to avoid recalculating forward and backward probabilities. The initial parameters are

$$\mathbf{A} = \begin{bmatrix} 0.7 & 0.3 \\ 0.2 & 0.8 \end{bmatrix}, \ \mathbf{B} = \begin{bmatrix} 0.75 & 0.25 \\ 0.40 & 0.60 \end{bmatrix}, \textbf{and } \mathbf{p} = (\mathbf{1}, \mathbf{0}).$$

We know from Example 1.2 that with the initial parameters, the probability of observations is $\mathbb{P}(O|\lambda) = \mathbf{0.11814}$. To update the model's parameters, we need to calculate the expected probabilities $\gamma_t(i)$ and $\xi_t(i,j)$ and use these values to re-estimate the model's parameters with the hope that the new parameter will increase the likelihood of the model. Using the forward and backward results from the previous examples we will calculate $\gamma_t(i)$ and $\xi_t(i,j)$ at time t.

At time $t = 1$:

$$\begin{aligned} \gamma_1(1) &= 1, \gamma_1(2) = 0 && (2.41) \\ \xi_1(1,1) &= 0.39438, \xi_1(1,2) = 0.60562 && (2.42) \\ \xi_1(2,1) &= 0, \xi_1(2,2) = 0. && (2.43) \end{aligned}$$

At time $t = 2$:

$$\begin{aligned} \gamma_2(1) &= 0.39438, \gamma_2(2) = 0.60562 && (2.44) \\ \xi_2(1,1) &= 0.19441, \xi_2(1,2) = 0.19997 && (2.45) \\ \xi_2(2,1) &= 0.05713, \xi_2(2,2) = 0.54848. && (2.46) \end{aligned}$$

At time $t = 3$:

$$\gamma_3(1) = 0.25155, \gamma_3(2) = 0.74845. \qquad (2.47)$$

Now, we use the numbers to re-estimate the model's parameters in the second interaction using the Baum-Welch algorithm 5:

- The probability of being in state S_i at time $t = 1$, $\mathbf{p}^* = (\gamma_1(\mathbf{1}), \gamma_1(\mathbf{2})) = (\mathbf{1}, \mathbf{0})$.
- The state transition probability $\mathbf{A}^*$:

$$\begin{aligned} a_{11}^* &= \frac{\xi_1(1,1)+\xi_2(1,1)}{\gamma_1(1)+\gamma_2(1)} = \frac{0.39438+0.19441}{1+0.39438} = 0.42226 \\ a_{12}^* &= \frac{\xi_1(1,2)+\xi_2(1,2)}{\gamma_1(1)+\gamma_2(1)} = \frac{0.60562+0.19997}{1+0.39438} = 0.57774 \\ a_{21}^* &= \frac{\xi_1(2,1)+\xi_2(2,1)}{\gamma_1(2)+\gamma_2(2)} = \frac{0+0.05713}{0+0.60562} = 0.09434 \\ a_{22}^* &= \frac{\xi_1(2,2)+\xi_2(2,2)}{\gamma_1(2)+\gamma_2(2)} = \frac{0+0.54848}{0+0.60562} = 0.90566. \end{aligned}$$

- The matrix of observation probability $\mathbf{B}^*$:

$$\begin{aligned} b_1^*(1) &= \frac{I_{\{O_1=Red\}}\gamma_1(1) + I_{\{O_2=Red\}}\gamma_2(1) + I_{\{O_3=Red\}}\gamma_3(1)}{\gamma_1(1)+\gamma_2(1)+\gamma_3(1)} \\ &= \frac{\gamma_1(1)}{\gamma_1(1)+\gamma_2(1)+\gamma_3(1)} \\ &= 0.60756 \\ b_1^*(2) &= \frac{I_{\{O_1=Blue\}}\gamma_1(1) + I_{\{O_2=Blue\}}\gamma_2(1) + I_{\{O_3=Blue\}}\gamma_3(1)}{\gamma_1(1)+\gamma_2(1)+\gamma_3(1)} \\ &= \frac{\gamma_2(1)+\gamma_3(1)}{\gamma_1(1)+\gamma_2(1)+\gamma_3(1)} \\ &= 0.39244 \\ b_2^*(1) &= \frac{I_{\{O_1=Red\}}\gamma_1(2) + I_{\{O_2=Red\}}\gamma_2(2) + I_{\{O_3=Red\}}\gamma_3(2)}{\gamma_1(1)+\gamma_2(1)+\gamma_3(1)} \\ &= \frac{\gamma_1(2)}{\gamma_1(1)+\gamma_2(1)+\gamma_3(1)} \\ &= 0 \\ b_2^*(2) &= \frac{I_{\{O_1=Blue\}}\gamma_1(2) + I_{\{O_2=Blue\}}\gamma_2(2) + I_{\{O_3=Blue\}}\gamma_3(2)}{\gamma_1(2)+\gamma_2(2)+\gamma_3(2)} \\ &= \frac{\gamma_2(2)+\gamma_3(2)}{\gamma_1(2)+\gamma_2(2)+\gamma_3(2)} \\ &= 1. \end{aligned}$$

In summary, we get the new parameters:

$$\mathbf{A}^* = \begin{bmatrix} 0.42226 & 0.57774 \\ 0.09434 & 0.90566 \end{bmatrix}, \ \mathbf{B}^* = \begin{bmatrix} 0.60756 & 0.39244 \\ 0.00000 & 1.0000 \end{bmatrix}, \ \textbf{and } \mathbf{p}^* = (\mathbf{1}, \mathbf{0}). \tag{2.48}$$

Using this fresh set of parameters, $\lambda^* = \{\mathbf{A}^*, \mathbf{B}^*, \mathbf{p}^*\}$, we apply the forward algorithm to calculate the new likelihood of the model, giving us the result

$$\mathbb{P}(O|\lambda^*) = \mathbf{0.40574}. \tag{2.49}$$

We see that after only one iteration, with the re-estimated parameters λ^*, the probability of observations increases from 0.11814 to 0.40574. If we continue the re-estimation process, we will have an optimal solution for the problem.

2.5.4 R Code for the Baum-Welch Algorithm

We will present R code to find the optimal parameters for a discrete HMM using the Baum-Welch algorithm. As shown in the algorithm, we need to calculate the forward probability α, backward probability β, and the conditional probability functions γ and ξ. We will write R functions to do the intermediate work and then call them in the Baum-Welch codes. We combine the forward and backward algorithms to one function to use it to calculate the γ and ξ functions. First, we present the R code for the forward-backward function:

```
# Dis.forward.back function calculates the forward and backward
# probabilities and the likelihood of a discrete HMM.
# Input:
#    x : observation sequence (length n)
#    A : transition matrix (m x m)
#    p : initial state probabilities (length m)
#    B : observation probability matrix (m x k)
# Output:
#    alpha : forward probabilities
#    beta  : backward probabilities
#    LL    : likelihood of the observation sequence
Dis.forward.back <- function(x, A, B, p) {
  m <- nrow(A)
  n <- length(x)
  # Forward pass
  phi <- as.double(p)
  alpha <- matrix(0, nrow = n, ncol = m)
  for (i in 1:n) {
    if (i > 1)
      phi <- phi %*% A
    phi <- phi * B[, x[i]]
    alpha[i, ] <- phi
  }
  ll <- sum(alpha[n, ])
  # Backward pass
  beta <- matrix(0, nrow = n, ncol = m)
  beta[n, ] <- rep(1, m)
  for (i in seq(n - 1, 1, -1))
    beta[i, ] <- A %*% (B[, x[i + 1]] * beta[i + 1, ])
  return(list(alpha = alpha, beta = beta, LL = ll))
}
```

Running the codes with the input from Example 1.2:

```
# Load the discrete forward-backward algorithm function
source("C:/Users/Dis.forward.back.R")
# Define transition and emission matrices
A <- matrix(c(0.7, 0.2,
              0.3, 0.8), ncol = 2, byrow = TRUE)
B <- matrix(c(0.75, 0.4,
              0.25, 0.6), ncol = 2, byrow = TRUE)
# Initial state probabilities and observation sequence
p <- c(1, 0)
x <- c(1, 2, 2)
```

```
# Compute forward and backward probabilities and likelihood
k <- Dis.forward.back(x, A, B, p)
print("The forward, backward probabilities, and the
likelihood are")
print(k)
```

we got the output results which include the vectors of forward and backward probabilities, and the likelihood of the observation $\mathbb{P}(O|\lambda)$. You can see that the results are the same as what we calculated by hand.

```
$alpha
           [,1]     [,2]
[1,] 0.75000000 0.000000
[2,] 0.13125000 0.135000
[3,] 0.02971875 0.088425

$beta
         [,1]    [,2]
[1,] 0.157525 0.27215
[2,] 0.355000 0.53000
[3,] 1.000000 1.00000

$LL
[1] 0.1181437
```

The forward-backward function for a continuous HMM is presented bellow.

```
# Cont.forward.back function calculates the forward and backward
# probabilities and the likelihood of a continuous HMM.
# Input:
#   x     : observation sequence (length n)
#   A     : transition matrix (m x m)
#   p     : initial state probabilities (length m)
#   mu    : vector of Gaussian means (length m)
#   sigma : vector of Gaussian standard deviations (length m)
# Output:
#   alpha : forward probabilities
#   beta  : backward probabilities
#   LL    : likelihood of the observation sequence
Cont.forward.back <- function(x, A, p, mu, sigma) {
  m <- nrow(A)
  n <- length(x)
  # Compute emission probabilities
  prob <- matrix(0, nrow = n, ncol = m)
  for (k in 1:m)
   prob[, k] <- dnorm(x, mean = mu[k], sd = sigma[k], log = FALSE)
  # Forward pass
  phi <- as.double(p)
  alpha <- matrix(0, nrow = n, ncol = m)
  for (i in 1:n) {
    if (i > 1)
      phi <- phi %*% A
```

```
    phi <- phi * prob[i, ]
    alpha[i, ] <- phi
  }
  ll <- sum(alpha[n, ])
  # Backward pass
  beta <- matrix(0, nrow = n, ncol = m)
  beta[n, ] <- rep(1, m)
  for (i in seq(n - 1, 1, -1))
    beta[i, ] <- A %*% (prob[i + 1, ] * beta[i + 1, ])
  return(list(alpha = alpha, beta = beta, LL = ll))
}
```

Applying the Cont.forward.back function to calculate the likelihood of observation from Example 1.6 from Chapter 1:

```
# Load the forward-backward algorithm for continuous HMM
source("C:/Users/Cont.forward.back.R")
# Define transition matrix
A <- matrix(c(0.4, 0.3, 0.3,
              0.2, 0.5, 0.3,
              0.6, 0.2, 0.2), ncol = 3, byrow = TRUE)
# Define Gaussian emission parameters
mu <- c(60, 55, 50)
sigma <- c(3, 2, 2.5)
# Initial state probabilities and observation sequence
p <- c(1, 0, 0)
x <- c(61, 57, 49)
# Compute forward and backward probabilities and likelihood
L <- Cont.forward.back(x, A, p, mu, sigma)
print("The forward backward probabilities, and the likelihood are")
print(L)
```

The output of the codes are:

```
$alpha
             [,1]         [,2]         [,3]
[1,] 1.257944e-01 0.000000e+00 0.0000000000
[2,] 4.058475e-03 4.565785e-03 0.0001194863
[3,] 4.175535e-07 7.809653e-06 0.0003846471

$beta
            [,1]        [,2]        [,3]
[1,] 0.003123146 0.003495416 0.003289833
[2,] 0.044921230 0.045332397 0.030000850
[3,] 1.000000000 1.000000000 1.000000000

$LL
[1] 0.0003928744
```

From the results, $LL = 0.0003928744$ is the likelihood, which is the same result that we found in the solution to Example 1.6. The next function is used to calculate the expectation probability γ defined in equation (2.18) and ξ defined in equation (2.22) of a discrete HMM.

```
\begin{lstlisting}[language=R]
# Dis.Estep function calculates expectation probabilities
# (Gamma and Xi) and the log-likelihood of observations
# for a discrete HMM.
# Input:
#    x : observation sequence (length n)
#    A : transition matrix (m x m)
#    p : initial state probabilities (length m)
#    B : observation probability matrix (m x k)
# Output:
#    u  : posterior state probabilities (Gamma)
#    v  : joint transition probabilities (Xi)
#    LL : total log-likelihood of the observation sequence
Dis.Estep <- function(x, A, B, p) {
  m <- nrow(A)
  n <- length(x)
  # Run forward-backward algorithm
  h <- Dis.forward.back(x, A, B, p)
  alpha <- h$alpha
  beta <- h$beta
  LL <- h$LL
  # Compute Gamma
  u <- alpha * beta / LL
  # Compute Xi
  v <- array(NA, dim = c(n - 1, m, m))
  for (k in 1:(n - 1)) {
    for (i in 1:m) {
      for (j in 1:m) {
        v[k, i, j] <- (beta[k + 1, j] * B[j, x[k + 1]]) *
                      (alpha[k, i] * A[i, j]) / LL
      }
    }
  }
  return(list(u = u, v = v, LL = LL))
}
```

Running the codes with the input from Example 1.2:

```
# Load the E-step function for discrete HMM
source("C:/Users/Dis.Estep.R")
# Define transition and emission matrices
A <- matrix(c(0.7, 0.2,
              0.3, 0.8), ncol = 2, byrow = TRUE)
B <- matrix(c(0.75, 0.4,
              0.25, 0.6), ncol = 2, byrow = TRUE)
# Initial state probabilities and observation sequence
p <- c(1, 0)
x <- c(1, 2, 2)
# Perform E-step: compute Gamma, Xi, and likelihood
GammaXi <- Dis.Estep(x, A, B, p)
print("Gamma, Xi, and likelihood are")
print(GammaXi)
```

we got identical results to what we calculated in (2.41)-(2.47).

```
$u (gamma)
          [,1]      [,2]
[1,] 1.0000000 0.0000000
[2,] 0.3943818 0.6056182
[3,] 0.2515474 0.7484526

$v (xi)
, , 1
          [,1]      [,2]
[1,] 0.3943818 0.6056182
[2,] 0.0000000 0.0000000
, , 2
          [,1]      [,2]
[1,] 0.19441359 0.1999683
[2,] 0.05713379 0.5484844
$LL
[1] 0.1181437
```

The next R function is the Dis.Mstep function that updates the matrix of observations **B** of a discrete HMM.

```
# Dis.Mstep function updates the observation probability matrix B
# for a discrete HMM using the posterior state probabilities.
# Input:
#   x    : observation sequence (length n)
#   cond : object returned from Dis.Estep containing gamma (u)
#   B    : initial observation probability matrix (m x k)
# Output:
#   B    : updated observation probability matrix
Dis.Mstep <- function(x, cond, B) {

  m <- ncol(B)
  for (k in 1:m) {
    y <- ifelse(x == k, 1, 0)
    B[, k] <- y %*% cond$u
  }
  B <- B / apply(cond$u, MARGIN = 2, FUN = sum)
  return(B)
}
```

Finally, we input all of these functions for a discrete HMM into the Dis.BaumWelch function below.

```
# Dis.BaumWelch function estimates parameters of a discrete HMM
# using the Baum-Welch algorithm.
# Input:
#   x        : observation sequence (length n)
#   A        : transition matrix (m x m)
#   p        : initial state probabilities (length m)
#   B        : observation probability matrix (m x k)
#   maxiter  : maximum number of iterations
```

```
#    tol          : convergence tolerance
# Output:
#    A list containing updated HMM parameters, expectations,
#    log-likelihood, convergence error, and iteration count
#Call the required functions
source("C:/Users/Dis.Estep.R")  # Expectation step
source("C:/Users/Dis.Mstep.R")  # Maximization step
Dis.BaumWelch <- function(x, A, B, p, maxiter, tol) {
  prt <- TRUE
  posdiff <- TRUE
  converge <- expression(diff < tol)
  m <- nrow(A)
  n <- length(x)
  oldLL <- -100
  for (iter in 1:maxiter) {
    cond <- Dis.Estep(x, A, B, p)
    diff <- cond$LL - oldLL
    # Update transition matrix
    A <- diag(1 / apply(cond$v, MARGIN = 2, FUN = sum)) %*%
         apply(cond$v, MARGIN = c(2, 3), FUN = sum)
    # Update initial state probabilities
    p <- cond$u[1, ]
    # Update observation probabilities
    B <- Dis.Mstep(x, cond, B)
    oldLL <- cond$LL
    if (diff < 0 && posdiff)
      stop("Worse likelihood on last iteration")
    if (eval(converge))
      break
  }
  object <- list(A = A, B = B, p = p,
                 u = cond$u, v = cond$v,
                 LL = cond$LL, diff = diff, iter = iter)
  return(object)
}
```

We will run the Dis.BaumWelch function with our initial parameters from Example 1.2. We chose the maximum number of interactions as $maxinter = 10$ and the tolerance as $tol = 1e-03$ for this example. Note that we need to list all functions that we used to calibrate the model's parameters.

```
# Load all required functions for Baum-Welch training
source("C:/Users/Dis.Estep.R")
source("C:/Users/Dis.forward.back.R")
source("C:/Users/Dis.BaumWelch.R")
source("C:/Users/Dis.Mstep.R")
# Define transition and emission matrices
A <- matrix(c(0.7, 0.2,
              0.3, 0.8), ncol = 2, byrow = TRUE)
B <- matrix(c(0.75, 0.4,
              0.25, 0.6), ncol = 2, byrow = TRUE)
# Initial state probabilities and observation sequence
```

```
p <- c(1, 0)
x <- c(1, 2, 2)
# Set training parameters
maxiter <- 10
tol <- 1e-03
# Run Baum-Welch algorithm
b <- Dis.BaumWelch(x, A, B, p, maxiter, tol)
print(b)
```

We got the results:

```
$A
             [,1] [,2]
[1,] 8.200033e-22    1
[2,] 1.672203e-22    1
$B
     [,1]         [,2]
[1,]    1 9.872236e-22
[2,]    0 1.000000e+00
$p
[1] 1 0
$u
             [,1] [,2]
[1,] 1.000000e+00    0
[2,] 8.200033e-22    1
[3,] 1.672203e-22    1
$v
, , 1
             [,1] [,2]
[1,] 8.200033e-22    1
[2,] 0.000000e+00    0
, , 2
             [,1]         [,2]
[1,] 6.724054e-43 8.200033e-22
[2,] 1.672203e-22 1.000000e+00
$LL
[1] 1
$diff
[1] 1.121059e-05
$iter
[1] 7
```

With the initial inputs, the Baum-Welch algorithm stopped after seven interactions with the error, the difference of the likelihoods between the two last interactions, as 0.00001. The calibrated parameters of the HMM are rounded to

$$\mathbf{A} = \begin{bmatrix} 0 & 1 \\ 0 & 1 \end{bmatrix}, \ \mathbf{B} = \begin{bmatrix} 1 & 0 \\ 0 & 1 \end{bmatrix}, \ \textbf{and } \mathbf{p} = (\mathbf{1}, \mathbf{0}).$$

With these parameters, the probability of observations is $\mathbb{P}(O|\lambda^*) = \mathbf{1}$. The likelihood is exactly one and larger than the likelihood when we calculated with the initial parameters, which was $\mathbb{P}(O|\lambda) = \mathbf{0.118144}$. These results mean that Box 1 should have all Red balls, and Box 2 should have all Blue balls. With the initial $\mathbf{p} = [1, 0]$, the first ball will be drawn from the Box 1 so the first ball would certainly be a Red ball. With the transition matrix $\mathbf{A}$, the second ball would be drawn from the Box 2, and then the third ball would be drawn from the Box 2. Therefore, the probability of observations $O = \{\text{Red, Blue, Blue}\}$ will certainly be one. After one iteration of the Dis.Baum-Welch algorithm, the results are

```
> A
            [,1]      [,2]
[1,] 0.42226269 0.5777373
[2,] 0.09433962 0.9056604
> B
           [,1]      [,2]
[1,] 0.6075595 0.3924405
[2,] 0.0000000 1.0000000
> p
[1] 1 0
> cond$LL
[1] 0.1181437
```

The results are consistent with those calculated manually in (2.48). Using the updated parameters, the likelihood of the observations is $\mathbb{P}(O|\lambda) = \mathbf{0.4057419}$, which also aligns with the result in (2.49).

2.5.5 Example of Using the Baum-Welch R Code

In this section, we will present three examples of "Weather Forecast" to enlighten how to use algorithms of the three main HMM problems to solve a real-life problem using both the handheld calculation and the R code.

Imagine that you are in a meeting room trying to figure out the weather outside (which is hidden from you) based on whether someone is carrying an umbrella or not (which you can observe) when they come to the meeting. Suppose that we have:

- Two hidden states: Sunny and Rainy
- Two possible observations (symbols): Umbrella, NO Umbrella
- State transition probability matrix $\mathbf{A}$:

	Sunny	Rainy
Sunny	0.8	0.2
Rainy	0.4	0.6

,

or

$$\mathbf{A} = \begin{bmatrix} 0.8 & 0.2 \\ 0.4 & 0.6 \end{bmatrix}. \tag{2.50}$$

That means if it's Sunny today, 80% chance it stays Sunny and 20% chance it changes to Rainy tomorrow. If it is Rainy today, 40% it will be Sunny tomorrow and 60% it will be Rainy tomorrow.

- Emission probabilities (B):

State	P(Umbrella)	P(No Umbrella)
Sunny	0.1	0.9
Rainy	0.8	0.2

,

or

$$\mathbf{B} = \begin{bmatrix} 0.1 & 0.9 \\ 0.8 & 0.2 \end{bmatrix}. \tag{2.51}$$

That means, on a Sunny day, only 10% chance the person carries umbrella, and on a Rainy day, 80% chance the person carries an umbrella.

- Initial probabilities $\mathbf{p}$:

$$\mathbf{p} = (0.5, 0.5).$$

That means we assume that the probability of Sunny or Rainy at time $t = 1$ is 50% for each state.

- Observation sequence O:

$$O = \{\text{Umbrella, Umbrella, No Umbrella}\}.$$

Thus, the parameters of the HMM for the "Weather Forecast" problem can be written as:

$$\lambda = \{\mathbf{A}, \mathbf{B}, \mathbf{p}\}. \tag{2.52}$$

We will present three examples that represent the three main problems of HMM based on the case of the "Weather Forecast".

Example 2.1 (Evaluation problem) *Calculate the probability of the observation sequence $O = \{$Umbrella, Umbrella, No Umbrella$\}$, given the HMM parameter λ.*

Solution 2.1
Note that the model parameter is $\lambda = \{\mathbf{A}, \mathbf{B}, \mathbf{p}\}$, number of states $N = 2, S_1 =$ Sunny, $S_2 =$ Rainy, and number of observations $T = 3$. By using the forward algorithm in Section 2.2.1, we have:

$$\mathbb{P}(O|\lambda) = \alpha_3(1) + \alpha_3(2) = \mathbf{P}(\mathbf{O_1}, \mathbf{O_2}, \mathbf{O_3}|\mathbf{q_3} = \mathbf{S_1}) + \mathbf{P}(\mathbf{O_1}, \mathbf{O_2}, \mathbf{O_3}|\mathbf{q_3} = \mathbf{S_2}),$$

where q_3 is the last state in the hidden state sequence $q = q_1, q_2, q_3$. Since $\mathbf{p} = (p_1, p_2) = (0.5, 0.5)$ at time t_1, we have:

$$\begin{aligned}
\alpha_1(1) &= \mathbb{P}(O_1|q_1 = S_1) = \mathbb{P}(Umbrella|q_1 = Sunny) \\
&= p_1 b_1(O_1) = (0.5)(0.1) \\
&= 0.05 \\
\alpha_1(2) &= \mathbb{P}(O_1|q_1 = S_2) = \mathbb{P}(Umbrella|q_1 = Rainy) \\
&= p_2 b_2(O_1) = (0.5)(0.8) \\
&= 0.4.
\end{aligned}$$

Using the recursive forward equation (2.2), at time t_2 we have:

$$\begin{aligned}
\alpha_2(1) &= [\alpha_1(1)a_{11} + \alpha_1(2)a_{21}]b_1(O_2) \\
&= [0.05(0.8) + 0.4(0.4)](0.1) \\
&= 0.02 \\
\alpha_2(2) &= [\alpha_1(1)a_{12} + \alpha_1(2)a_{22}]b_2(O_2) \\
&= [0.05(0.2) + 0.4(0.6)](0.8) \\
&= 0.2.
\end{aligned}$$

Similarly, at time t_3 we have

$$\begin{aligned}\alpha_3(1) &= [\alpha_2(1)a_{11} + \alpha_2(2)a_{21}]b_1(O_3)\\ &= [0.02(0.8) + 0.2(0.4)](0.9)\\ &= 0.0864\\ \alpha_3(2) &= [\alpha_2(1)a_{12} + \alpha_2(2)a_{22}]b_2(O_3)\\ &= [0.02(0.2) + 0.2(0.6)](0.2)\\ &= 0.0248.\end{aligned}$$

Thus, the probability of observation is

$$\mathbb{P}(O|\lambda) = \alpha_3(1) + \alpha_3(2) = \mathbf{0.0864 + 0.0248 = 0.1112}.$$

Now we will use the forward function of a discrete HMM (Dis.forward) R code function in Section 2.2.4 to calculate the observation probability. First, the observation sequence was transformed to a vector, **x**, of binary values based on the two symbols: Umbrella or No Umbrella.

```
# Load the discrete forward algorithm function
source("C:/Users/Dis.forward.R")
# Define observation sequence and convert to numeric states
observations <- c("Umbrella", "Umbrella", "No Umbrella")
x <- ifelse(observations == "Umbrella", 1, 2)
# Define initial state probabilities
p <- c(0.5, 0.5)
# Define transition and emission matrices
A <- matrix(c(0.8, 0.2,
              0.4, 0.6), nrow = 2, byrow = TRUE)
B <- matrix(c(0.1, 0.9,
              0.8, 0.2), nrow = 2, byrow = TRUE)
# Compute likelihood using forward algorithm
L <- Dis.forward(x, A, B, p)
cat("P(O | lambda) =", L, "\n")
```

The output of the codes is:

```
P(O | lambda) = 0.1112
```

We got the same result as using the manual calculations in Example 2.1.

Example 2.2 (Decoding problem) *In the "Weather Forecast" problem, given the parameter λ, and the observations $O = \{Umbrella, Umbrella, No\ Umbrella\}$., find the best hidden state sequence $Q^* = \{q_1.q_2, q_3\}$ that maximizes the probability of observations $P(O|\lambda)$.*

Solution 2.2

To find the optimal hidden sequence Q^* using the Viterbi algorithm in Section 2.4, we will calculate values of the probability function $\delta_t(j)$ and the index function $\Phi_t(j)$ at each time t recursively. From Example 2.1 we first transfer the observation sequence to a vector of binary value based on the two symbols: O = {Umbrella, Umbrella, No Umbrella} $= \{\nu_1, \nu_1, \nu_2\} = \{1, 1, 2\}$. At time $t = 1$, we have:

$$\begin{aligned}\delta_1(1) &= p_1b_1(O_1) = p_1b_1(1) = p_1b_{11} = 0.5(0.1) = 0.05\\ \delta_1(2) &= p_2b_2(O_1) = p_2b_2(1) = p_2b_{21} = 0.5(0.8) = 0.40\\ \Phi_1(1) &= \Phi_1(2) = 0.\end{aligned}$$

At time $t = 2$, we have:

$$\begin{aligned}
\delta_2(1) &= b_1(O_2)\max_{i=1,2}\delta_1(i)a_{i1} = 0.1\max\{0.05(0.8), 0.4(0.4)\}\\
&= 0.16(0.1) = 0.016\\
\delta_2(2) &= b_2(O_2)\max_{i=1,2}\{\delta_1(i)a_{i2}\} = 0.8\max\{\ 0.05(0.2), 0.4(0.6)\}\\
&= 0.24(0.8) = 0.192\\
\Phi_2(1) &= \arg\max_{i=1,2}\{\delta_1(i)a_{i1}\} = \arg\max_{i=1,2}\{0.05(0.8), 0.4(0.4)\} = 2\\
\Phi_2(2) &= \arg\max_{i=1,2}\{\delta_1(i)a_{i2}\} = \arg\max_{i=1,2}\{\ 0.05(0.2), 0.4(0.6)\} = 2.
\end{aligned}$$

At time $t = 3$, we have:

$$\begin{aligned}
\delta_3(1) &= b_1(O_3)\max_{i=1,2}\{\delta_2(i)a_{i1}\} = 0.9\max\{0.016(0.8), 0.192(0.4)\}\\
&= 0.9\max\{0.0128, 0.0768\} = 0.9(0.0768) = 0.06912\\
\delta_3(2) &= b_2(O_3)\max_{i=1,2}\{\delta_2(i)a_{i2}\} = 0.2\max\{0.016(0.2), 0.192(0.6)\}\\
&= 0.2\max\{0.0032, 0.1152\} = 0.2(0.1152) = 0.02304\\
\Phi_3(1) &= \arg\max_{i=1,2}\{\delta_2(i)a_{i1}\} = \arg\max_{i=1,2}\{0.016(0.8), 0.192(0.4)\} = 2\\
\Phi_3(2) &= \arg\max_{i=1,2}\{\delta_2(i)a_{i2}\} = \arg\max_{i=1,2}\{0.016(0.2), 0.192(0.6)\} = 2.
\end{aligned}$$

Thus, the "best fit" hidden states are:

$$\begin{aligned}
q_3^* &= \arg\max_{i=1,2}\{\delta_3(i)\} = \arg\max_{i=1,2}\{0.06912, 0.02304\} = 1\\
q_2^* &= \Phi_3(q_3^*) = \Phi_3(1) = 2\\
q_1^* &= \Phi_2(q_2^*) = \Phi_2(2) = 2.
\end{aligned}$$

In summary, the "best fit" state sequence is

$$Q^* = \{q_1^*, q_2^*, q_3^*\} = \{2, 2, 1\} = \{S_2, S_2, S_1\} = \{\text{Rainy,Rainy,Sunny}\}.$$

We then use the R code of the Viterbi algorithms for a discrete HMM (Dis.Viterbi) in Section 2.4.3 to find the hidden state Q^*.

```
# Load the discrete Viterbi algorithm function
source("C:/Users/Dis.Viterbi.R")
# Define symbolic observation sequence and convert to numeric
states observations <- c("Umbrella", "Umbrella", "No Umbrella")
x <- ifelse(observations == "Umbrella", 1, 2)
# Define initial state probabilities
p <- c(0.5, 0.5)
# Define transition and emission matrices
A <- matrix(c(0.8, 0.2,
              0.4, 0.6), nrow = 2, byrow = TRUE)
B <- matrix(c(0.1, 0.9,
              0.8, 0.2), nrow = 2, byrow = TRUE)
# Run Viterbi algorithm to find most likely hidden state sequence
S <- Dis.Viterbi(A, B, p, x)
print("Hidden state sequence is Q =")
print(S)
```

The output hidden state is $Q = \{2, 2, 1\}$ or equivalent $Q = \{\text{Rainy, Rainy, Sunny}\}$. This result agrees with the result that we found by manual calculations. We will move to the last example with the "Weather Forecast" problem.

Example 2.3 (Learning problem) *In the "Weather Forecast" problem, given the observations* $O = \{Umbrella, Umbrella, NoUmbrella\}$, *estimate the model parameters* $\lambda = \{\mathbf{A}, \mathbf{B}, \mathbf{p}\}$.

Solution 2.3
We will use Baum-Welch algorithm to solve this problem. A similar example was presented in Section 2.5.3. The algorithm involves many iterations until it reaches the maximum number of the iterations or the designed error. Therefore, we will not present the manual calculations. Readers can find similar manual calculation steps in the example in Section 2.5.3. We will use the Baum.Welch algorithm R code to find the model parameters. First, we choose the initial parameters:

$$\mathbf{p} = (0.5, 0.5).$$

Transition matrix **A**

	Sunny	Rainy
Sunny	0.6	0.4
Rainy	0.5	0.5

or

$$\mathbf{A} = \begin{bmatrix} 0.6 & 0.4 \\ 0.5 & 0.5 \end{bmatrix}. \tag{2.53}$$

Emission matrix **B**

State	P(Umbrella)	P(No Umbrella)
Sunny	0.7	0.3
Rainy	0.4	0.6

or

$$\mathbf{B} = \begin{bmatrix} 0.7 & 0.3 \\ 0.4 & 0.6 \end{bmatrix}. \tag{2.54}$$

We choose the maximum number of iterations is 20, ($maxiter = 20$), and the error is 0.003, ($tol = 10^{-3}$).

```
# Load all required functions for Baum-Welch training
source("C:/Users/Dis.Estep.R")
source("C:/Users/Dis.forward.back.R")
source("C:/Users/Dis.BaumWelch.R")
source("C:/Users/Dis.Mstep.R")
# Define transition and emission matrices
A <- matrix(c(0.6, 0.5,
              0.4, 0.5), ncol = 2, byrow = TRUE)
B <- matrix(c(0.7, 0.4,
              0.3, 0.6), ncol = 2, byrow = TRUE)
# Initial state probabilities and symbolic observation sequence
p <- c(0.5, 0.5)
observations <- c("Umbrella", "Umbrella", "No Umbrella")
x <- ifelse(observations == "Umbrella", 1, 2)
# Set training parameters
maxiter <- 10
tol <- 1e-03
# Run Baum-Welch algorithm
```

```
b <- Dis.BaumWelch(x, A, B, p, maxiter, tol)
print(b)
```

The outputs of running the codes are:

```
$A
           [,1]      [,2]
[1,] 0.7520321 0.2479679
[2,] 0.1561330 0.8438670
$B
           [,1]      [,2]
[1,] 0.8671427 0.1328573
[2,] 0.1561830 0.8438170
$p
[1] 9.999498e-01 5.018327e-05
$u
           [,1]         [,2]
[1,] 0.9999498 5.018327e-05
[2,] 0.8679290 1.320710e-01
[3,] 0.2861827 7.138173e-01
$v
, , 1
              [,1]         [,2]
[1,] 8.679133e-01 1.320365e-01
[2,] 1.560788e-05 3.457539e-05
, , 2
             [,1]      [,2]
[1,] 0.281958036 0.5859709
[2,] 0.004224654 0.1278464
$LL
[1] 0.1995292
$diff
[1] 0.0001016215
$iter
[1] 9
```

The results showed that the Baum-Welch algorithm stopped after nine iterations and had the error of 0.000101, which is less than the defined error of 0.001. The results displayed the transition matrix **A**, the emission matrix **B**, the vector initial probabilities **p**. In a continuous HMM, we typically deal with a large number of observations. When using the original algorithms to calculate the probability of these observations, the probability is often a very small non-negative number, which may exceed the computer's tolerance for precision. In the next section, we address this issue and provide scaling R code for a continuous HMM, along with practical applications.

2.6 Scaling HMM Algorithms with R Code

In this section, we address a practical challenge in implementing algorithms for HMMs. While issues such as handling missing data or managing large datasets fall outside the scope of this discussion, we focus on the practical aspects of HMM algorithms and their implementation in R. Among these algorithms, the Baum-Welch algorithm stands out as particularly important because it is commonly applied in real-world scenarios where only observation data is available. This algorithm estimates the parameters of an HMM by utilizing both the forward and backward algorithms to compute observation probabilities. To provide a foundation, we start with the forward algorithm and illustrate the scaling method introduced in [115].

We can see from Section 2.2 that all of the HMM algorithms involve calculating the probability of the observations, $\mathbb{P}(O|\lambda)$. We use the forward and backward algorithms to calculate the likelihood. However, the forward function $\alpha_t(i)$ at time t, presented in equations (2.1) and (2.2), is the sum of the products of a large number of terms, each of the form

$$\prod_{s=1}^{t-1} a_{q_s q_{s+1}} \prod_{s=1}^{t} b_{q_s}(O_s), \tag{2.55}$$

where $q_t = S_i$. Since the transition probability $a_{q_s q_{s+1}}$ and the observation probability $b_{q_s}(O_s)$ are non-negative numbers and less than one, the product of the terms (2.55) will exponentially approach zero. Therefore, when t becomes large, (e.g. $t > 100$), the range of the product computation may exceed the numerical precision limits of most machines. Thus, we must scale the numbers to make the calculations work. The basic idea of scaling is to multiply the "too small" term $\alpha_t(i)$ by a scaling coefficient, that is independent from the variable i, to make the scaled $\alpha_t(i)$ fall within the computational range of a computer.

In the forward algorithm, we will scale the forward function $\alpha_t(i)$ by multiplying it by the coefficient

$$c_t = \frac{1}{\sum_{i=1}^{N} \alpha_t(i)}. \tag{2.56}$$

We denote $\hat{\alpha}_t(i)$ as the scaled function of $\alpha_t(i)$. Therefore, at any time t, the scaled forward function has the property

$$\sum_{i=1}^{N} \hat{\alpha}_t(i) = c_t(\sum_{i=1}^{N} \alpha_t(i)) = 1. \tag{2.57}$$

Next, we will find a recursive formula for $\hat{\alpha}_t(i)$. At time $t = 1$, we have

$$\hat{\alpha}_1(i) = c_1\alpha_1(i). \tag{2.58}$$

At time $t \geq 2$, we first calculate $\alpha_t(i)$ by using the forward equation (2.2):

$$\alpha_t(i) = \sum_{j=1}^{N} \hat{\alpha}_{t-1}(j) a_{ij} b_j(O_t). \tag{2.59}$$

Then, calculate the scaled $\hat{\alpha}_t(i)$

$$\begin{aligned} \hat{\alpha}_t(i) &= c_t\alpha_t(i) && (2.60) \\ &= \frac{\sum_{j=1}^{N} \hat{\alpha}_{t-1}(j) a_{ij} b_j(O_t)}{\sum_{i=1}^{N}\sum_{j=1}^{N} \hat{\alpha}_{t-1}(j) a_{ij} b_j(O_t)}. && (2.61) \end{aligned}$$

By induction, we have

$$\hat{\alpha}_t(j) = (\prod_{\tau=1}^{t} c_\tau)\alpha_t(j), 1 \leq t \leq T, \tag{2.62}$$

where c_τ is defined in (2.56). In the forward algorithm from Section 1, the probability of observation is calculated by

$$\mathbb{P}(O|\lambda) = \sum_{\mathbf{i=1}}^{\mathbf{N}} \alpha_{\mathbf{T}}(\mathbf{i}).$$

Note that in the scaled forward algorithm we will have the scaled $\hat{\alpha}_t$, not α_t. Thus, in the scaled forward algorithm we have

$$\hat{\alpha}_T(i) = \prod_{\tau=1}^{T} c_\tau \alpha_T(i). \tag{2.63}$$

Applying the property (2.57) for the scaled function, we have

$$1 = \sum_{i=1}^{N} \hat{\alpha}_T(i) = (\prod_{\tau=1}^{T} c_\tau) \cdot \sum_{i=1}^{N} \alpha_T(i) = (\prod_{\tau=1}^{T} c_\tau) \cdot \mathbb{P}(O|\lambda). \tag{2.64}$$

Finally, we have the likelihood of the observations

$$\mathbb{P}(O|\lambda) = \frac{\mathbf{1}}{\prod_{\tau=1}^{\mathbf{T}} \mathbf{c}_\tau}. \tag{2.65}$$

In practice, $\mathbb{P}(O|\lambda)$ will exponentially approach zero when the number of observation T is big. Therefore, we will calculate the logarithm of the probability,

$$\log \mathbb{P}(O|\lambda) = \log(\frac{\mathbf{1}}{\prod_{\tau=1}^{\mathbf{T}} \mathbf{c}_\tau}) = -\log(\prod_{\tau=1}^{\mathbf{T}} \mathbf{c}_\tau) = -(\sum_{\tau=1}^{\mathbf{T}} \log \mathbf{c}_\tau). \tag{2.66}$$

In the next section, we will present R code for a continuous HMM, assuming with assumption that the probability observations in each state follow a Gaussian distribution with the mean μ and standard deviation σ.

2.6.1 Scaled Forward Algorithm

The R code for the forward function of a continuous HMM using the scaling method are presented below. In the function, we calculate the likelihood by

$$\log \mathbb{P}(O|\lambda) = \log(\frac{\mathbf{1}}{\prod_{\tau=1}^{\mathbf{T}} \mathbf{c}_\tau}) = \sum_{\tau=1}^{\mathbf{T}} \log(\frac{\mathbf{1}}{\mathbf{c}_\tau}) = \sum_{\tau=1}^{\mathbf{T}} \log(\sum_{\mathbf{i=1}}^{\mathbf{N}} \alpha_{\mathbf{t}}(\mathbf{i})). \tag{2.67}$$

```
# scaled.forward function calculates the likelihood of a
# continuous HMM using the scaled forward algorithm.
# Input:
#   x     : observation sequence (length n)
#   A     : transition matrix (m x m)
#   p     : initial state probabilities (length m)
#   mu    : vector of means of Gaussian distributions (length m)
#   sigma : vector of standard deviations of Gaussian
#           distributions (length m)
# Output:
#   LL    : likelihood of the observation sequence
scaled.forward <- function(x, A, p, mu, sigma) {
```

```
  m <- nrow(A)
  n <- length(x)
  # Compute emission probabilities
  prob <- matrix(0, nrow = n, ncol = m)
  for (k in 1:m)
   prob[, k] <- dnorm(x, mean = mu[k], sd = sigma[k], log = FALSE)
  # Scaled forward recursion
  phi <- as.double(p)
  llike <- 0
  for (i in 1:n) {
    if (i > 1)
      phi <- phi %*% A
    phi <- phi * prob[i, ]
    sumphi <- sum(phi)
    phi <- phi / sumphi
    llike <- llike + log(sumphi)
  }
  # Final likelihood
  LL <- exp(llike)
  return(LL)
}
```

Using the scaled.forward function codes to solve Example 1.6 from Chapter 1,

```
# Load the scaled forward algorithm function
source("C:/Users/scaled.forward.R")
# Define transition matrix
A <- matrix(c(0.4, 0.3, 0.3,
              0.2, 0.5, 0.3,
              0.6, 0.2, 0.2), ncol = 3, byrow = TRUE)
# Define Gaussian emission parameters
mu <- c(60, 55, 50)
sigma <- c(3, 2, 2.5)
# Initial state probabilities and observation sequence
p <- c(1, 0, 0)
x <- c(61, 57, 49)
# Compute likelihood using scaled forward algorithm
L <- scaled.forward(x, A, p, mu, sigma)
# Display result with formatted precision
sprintf("%#.6f is the likelihood of observations using scaled.
forward function", L)
```

we got the result: "0.000393 is the likelihood of observations using scaled.forward function". The result is the same as what we got in Example 1.6 from Chapter 1.

2.6.2 Scaled Backward Algorithm

We will do similar scaling for the backward function because its value will goes to zero exponentially when t is big. We use the same scale factors in the forward algorithm for the backward algorithm.

$$\hat{\beta}_t(i) = c_t \beta_t(i), 1 \leq i \leq N. \tag{2.68}$$

Similar to the scaled forward function, by induction we have

$$\hat{\beta}_{t+1}(i) = (\prod_{\tau=t+1}^{T} c_\tau)\beta_{t+1}(i). \tag{2.69}$$

The code using backward algorithm to calculate the probability of observations for a given continuous HMM.

```
# scaled.backward function calculates the likelihood of
# a continuous HMM
# using the scaled backward algorithm.
# Input:
#   x       : observation sequence (length n)
#   A       : transition matrix (m x m)
#   p       : initial state probabilities (length m)
#   mu      : vector of Gaussian means (length m)
#   sigma   : vector of Gaussian standard deviations (length m)

# Output:
#   ll      : likelihood of the observation sequence
scaled.backward <- function(x, A, p, mu, sigma) {
  m <- nrow(A)
  n <- length(x)
  # Compute emission probabilities
  prob <- matrix(0, nrow = n, ncol = m)
  for (k in 1:m)
   prob[, k] <- dnorm(x, mean = mu[k], sd = sigma[k], log = FALSE)
  # Scaled backward recursion
  phi <- rep(1 / m, m)
  llike <- log(m)
  for (i in seq(n - 1, 1, -1)) {
    phi <- A %*% (prob[i + 1, ] * phi)
    sumphi <- sum(phi)
    phi <- phi / sumphi
    llike <- llike + log(sumphi)
  }
  # Final likelihood computation
  ll <- sum(p * prob[1, ] * phi)
  return(ll)
}
```

Using the scaled.backward function to solve Example 1.6 from Chapter 1,

```
# Load the scaled backward algorithm function
source("C:/Users/scaled.backward.R")
# Define transition matrix
A <- matrix(c(0.4, 0.3, 0.3,
              0.2, 0.5, 0.3,
              0.6, 0.2, 0.2), ncol = 3, byrow = TRUE)
# Define Gaussian emission parameters
mu <- c(60, 55, 50)
sigma <- c(3, 2, 2.5)
```

```
# Initial state probabilities and observation sequence
p <- c(1, 0, 0)
x <- c(61, 57, 49)
# Compute likelihood using scaled backward algorithm
L <- scaled.backward(x, A, p, mu, sigma)
# Display result with formatted precision
sprintf("%#.6f is the likelihood of observations using scaled.
backward function.", L)
```

we got the result: "0.000393 is the likelihood of observations using scaled.backward function". The result is the same as what we got in Example 1.6 from Chapter 1.

Next, we will present another continuous HMM and apply the scaled.forward and scaled.backward functions to calculate the likelihood of observations.

Example 2.4 *Suppose that fifty observations from a mixed normal distribution which include the two different normal distributions with the means of* -1.1 *and* 1.5 *and standard deviations of* 0.35 *and* 0.5*, respectively. We assume that the observations follow a two state continuous HMM with the transition matrix*

$$\mathbf{A} = \begin{bmatrix} 0.7 & 0.3 \\ 0.2 & 0.8 \end{bmatrix}, \tag{2.70}$$

and the initial probability

$$\mathbf{p} = (0.5, 0.5).$$

Calculate the likelihood of the observations using the scaled.forward and scaled.backward algorithm functions.

Solution 2.4
We use the R code below to solve the problem.

```
# Call the scaled.forward and scaled.backward functions
source("C:/Users/scaled.forward.R")
source("C:/Users/scaled.backward.R")
# Initialize model parameters
A <- matrix(c(0.7, 0.2,
              0.3, 0.8), ncol = 2, byrow = TRUE)
mu <- c(-1.1, 1.5)
sigma <- c(0.35, 1.1)
p <- c(0.5, 0.5)
# Randomly generate 50 observations from two normal distributions
N <- 50
y <- floor(runif(N) * 2 + 1)  # Random states: 1 or 2
x <- numeric(N)
# Generate observations based on hidden states
for (i in 1:N) {
  if (y[i] == 1)
    x[i] <- rnorm(1, mean = mu[1], sd = sigma[1])
  else
    x[i] <- rnorm(1, mean = mu[2], sd = sigma[2])
}
# Calculate likelihood of observations
f <- scaled.forward(x, A, p, mu, sigma)
```

```
b <- scaled.backward(x, A, p, mu, sigma)
print("The likelihood of observations using scaled.forward is")
print(f)
print("The likelihood of observations using scaled.backward is")
print(b)
```

Output of running the codes are:

```
 "The likelihood of observations using scaled.forward function is"
> print(f)
 1.207704e-37
 "The likelihood of observations using scaled.backward function is"
> print(b)
 1.207704e-37.
```

We observe that the probability of fifty observations is extremely small, and the scaled.forward and scaled.backward functions yield the same result. Note that, since the observations are generated randomly, running the code multiple times will produce different results corresponding to the generated sequences.

2.6.3 Log Viterbi Algorithm and Its Example

We can see from the Viterbi algorithm in Section 2.4 that there is one term that needs scale, which is

$$\delta_t(j) = \max_{q_1, q_2, \ldots, q_{t-1}} \log \mathbb{P}(q_1, q_2, \ldots, q_t = S_j, O_1, \ldots, O_t|\lambda). \tag{2.71}$$

The initial value of the $\delta_t(j)$ at time $t = 1$ is

$$\delta_1(j) = \log(p_j b_j(O_1)) = \log(p_j) + \log(b_j(O_1)), \tag{2.72}$$

and the recursive function should be

$$\delta_{t+1}(j) = \log(b_j(O_{t+1}) + \max_i\{(\delta_t(i) + \log(a_{ij})\}. \tag{2.73}$$

The log of probability of observation in the hidden state is

$$log(\mathbb{P}(O, Q^*|\lambda)) = \max_{\mathbf{1 \le j \le N}} \delta_{\mathbf{T}}(\mathbf{i}).$$

The functions $\phi_t(j) = \arg\max_i[\delta_{t-1}(i)a_{ij}]$ and $q_T^* = \arg\max_i[\delta_T(i)]$ does not change. Thus, the output states are the same as in the original Viterbi algorithm in Section 2.4. Below is the Log.Viterbi R function for a continuous HMM.

```
# Log.Viterbi function computes the most likely hidden state
# sequence
# for a continuous HMM using the log-space Viterbi algorithm.
# Assumes HMM has two or more states.
# Inputs:
#   x       : observation sequence
#   A       : transition matrix between states
#   p       : initial state probabilities at time t = 1
#   mu      : vector of Gaussian means
#   sigma   : vector of Gaussian standard deviations
# Output:
```

```
#   Q       : most likely hidden state sequence
Log.Viterbi <- function(A, mu, sigma, p, x) {
  m <- nrow(A)
  T <- length(x)
  Q <- integer(T)
  theta <- matrix(0, T, m)
  Phi <- matrix(0, T, m)
  # Compute emission probabilities
  prob <- matrix(0, nrow = T, ncol = m)
  for (k in 1:m)
   prob[, k] <- dnorm(x, mean = mu[k], sd = sigma[k], log = FALSE)
  # Initialization
  theta[1, ] <- log(p) + log(prob[1, ])
  # Recursion
  for (t in 2:T) {
    for (i in 1:m) {
      scores <- theta[t - 1, ] + log(A[, i])
      theta[t, i] <- log(prob[t, i]) + max(scores)
      Phi[t, i] <- which.max(scores)
    }
  }
  # Backtracking
  Q[T] <- which.max(theta[T, ])
  for (t in (T - 1):1)
    Q[t] <- Phi[t + 1, Q[t + 1]]
  return(Q)
}
```

We will apply the Log.Viterbi function to find hidden state sequence of the 50 observations from a mix normal distributions in Example 2.4.

```
# Call the Log.Viterbi function
source("C:/Users/Log.Viterbi.R")
# Initialize model parameters
A <- matrix(c(0.7, 0.2,
              0.3, 0.8), ncol = 2, byrow = TRUE)
mu <- c(-1.1, 1.5)
sigma <- c(0.35, 1.1)
p <- c(0.5, 0.5)
# Randomly generate 50 observations from two normal distributions
N <- 50
y <- floor(runif(N) * 2 + 1)  # Random states: 1 or 2
x <- numeric(N)
# Generate observations based on hidden states
for (i in 1:N) {
  if (y[i] == 1)
    x[i] <- rnorm(1, mean = mu[1], sd = sigma[1])
  else
    x[i] <- rnorm(1, mean = mu[2], sd = sigma[2])
}
# Run log-space Viterbi algorithm to infer hidden states
k <- Log.Viterbi(A, mu, sigma, p, x)
```

```
# Plot predicted states vs actual states
t <- 1:N
matplot(t, cbind(y, k), type = "l",
        col = c("red", "blue"), xlab = "t", ylab = "States",
        main = "Actual States vs. Predicted States")
legend("topright", legend = c("Actual States", "Predicted States"),
       col = c("red", "blue"), lty = 1)
# Calculate prediction accuracy
confusion <- table(y, k)
print(confusion)
accuracy <- sum(diag(confusion)) / sum(confusion)
print("Accuracy of the state prediction is")
print(paste(round(100 * accuracy, 2), "%"))
```

The Log.Viterbi function identified the hidden states of the HMM with a high degree of accuracy (approximately 92% to 100%) using the initial parameter set. The output from running the code is shown in Figure (2.4). Note that, since a random function is used to generate observation sequences, the outcomes will vary each time the code is executed.

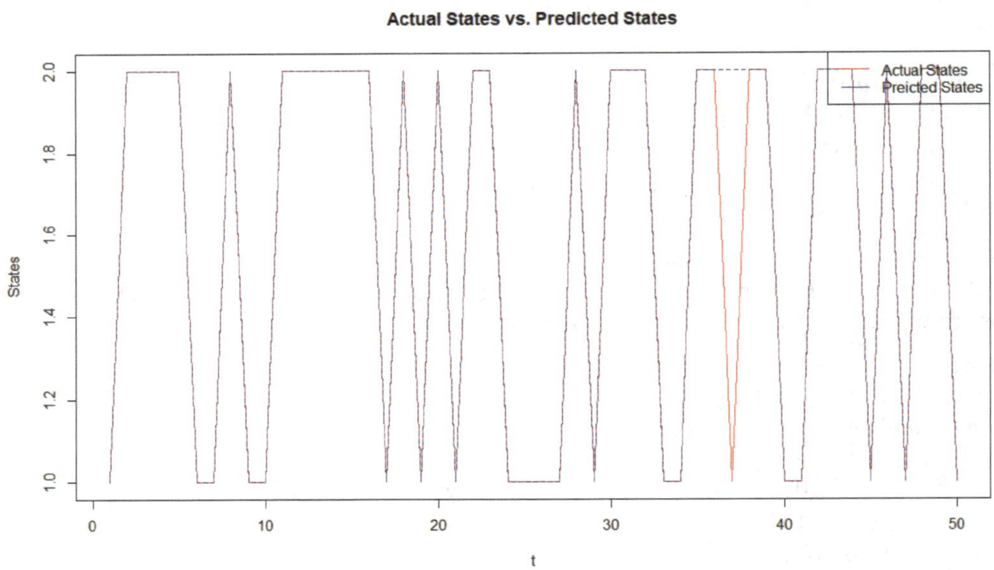

FIGURE 2.4
The actual states vs. predicted states of a continuous HMM using the Log.Viterbi algorithm.

The accuracy of the prediction is 98% as shown in the output

```
       Cross table of actual states vs, predicted states
             Predicted
Actual         1  2
   1          20  1
   2           0 29

"Accuracy of the state prediction is"
"98 %"
```

From the cross table and Figure 2.4, we see that the Log.Viterbi function only makes one wrong prediction of state 1.

We also can test the Log.Viterbi function by applying it to solve Example 1.7 in Chapter 1.

```
# Call the Log.Viterbi function
source("C:/Users/Log.Viterbi.R")
# Define transition matrix
A <- matrix(c(0.4, 0.3, 0.3,
              0.2, 0.5, 0.3,
              0.6, 0.2, 0.2), ncol = 3, byrow = TRUE)
# Define Gaussian emission parameters
mu <- c(60, 55, 50)
sigma <- c(3, 2, 2.5)
# Initial state probabilities and observation sequence
p <- c(1, 0, 0)
x <- c(61, 57, 49)
# Run log-space Viterbi algorithm to infer hidden state sequence
Q <- Log.Viterbi(A, mu, sigma, p, x)
print("The hidden state sequence is")
print(Q)
```

The output of the codes is: "The hidden state sequence is" Q={1,2,3}. The result means the hidden sequence is $Q = \{\text{Sunny, Rainy, Cloudy}\}$, which is identical to the solution of Example 1.7 in Chapter 1.

2.6.4 Scaled Baum-Welch Algorithm and Its Example

Since we use both forward and backward algorithms in the Baum-Welch algorithm, we will explore how the scaling method works in the algorithm.

When applying the scaling method to the Baum-Welch algorithm 5, the scaled transition matrix **A** is estimated by

$$\hat{a}^*_{ij} = \frac{\sum_{t=1}^{T-1} \xi_t(i,j)}{\sum_{t=1}^{T-1} \gamma_t(i)} \tag{2.74}$$

$$= \frac{\hat{\alpha}_t(i) a_{ij} b_j(O_{t+1}) \hat{\beta}_{t+1}(j)}{\sum_{j=1}^{N} \hat{\alpha}_t(i) a_{ij} b_j(O_{t+1}) \hat{\beta}_{t+1}(j)}. \tag{2.75}$$

Using the formulas for the scaled alpha (2.63) and scaled beta (2.69), we have

$$\hat{\alpha}_t(i) = (\prod_{\tau=1}^{t} c_\tau)\alpha_t(i) = C_t \alpha_t(i), \tag{2.76}$$

where $C_t = \prod_{\tau=1}^{t} c_\tau$.

$$\hat{\beta}_{t+1}(i) = (\prod_{\tau=t+1}^{T} c_\tau)\beta_{t+1}(i) = D_{t+1}\beta_{t+1}(i), \tag{2.77}$$

where $D_{t+1} = \prod_{\tau=t+1}^{T} c_\tau$.

We notice that

$$C_tD_{t+1} = \prod_{\tau=1}^{t} c_\tau \prod_{\tau=t+1}^{T} c_\tau = \prod_{\tau=1}^{T} c_\tau = C_T. \tag{2.78}$$

Thus, the product C_tD_{t+1} does not depend on t. If we substitute the terms in (2.76, 2.77, 2.78) into the equation (2.75), we have

$$\begin{aligned}
\hat{a}^*_{ij} &= \frac{\hat{\alpha}_t(i)a_{ij}b_j(O_{t+1})\hat{\beta}_{t+1}(j)}{\sum_{j=1}^{N}\hat{\alpha}_t(i)a_{ij}b_j(O_{t+1})\hat{\beta}_{t+1}(j)}. \\
&= \frac{C_t\alpha_t(i)a_{ij}b_j(O_{t+1})D_{t+1}\beta_{t+1}(j)}{\sum_{j=1}^{N}C_t\alpha_t(i)a_{ij}b_j(O_{t+1})D_{t+1}\beta_{t+1}(j)}. \\
&= \frac{C_tD_{t+1}\alpha_t(i)a_{ij}b_j(O_{t+1})\beta_{t+1}(j)}{\sum_{j=1}^{N}C_tD_{t+1}\alpha_t(i)a_{ij}b_j(O_{t+1})D_{t+1}\beta_{t+1}(j)}. \\
&= \frac{C_T\alpha_t(i)a_{ij}b_j(O_{t+1})\beta_{t+1}(j)}{\sum_{j=1}^{N}C_T\alpha_t(i)a_{ij}b_j(O_{t+1})D_{t+1}\beta_{t+1}(j)}. \\
&= \frac{\alpha_t(i)a_{ij}b_j(O_{t+1})\beta_{t+1}(j)}{\sum_{j=1}^{N}\alpha_t(i)a_{ij}b_j(O_{t+1})D_{t+1}\beta_{t+1}(j)}. \\
&= a^*_{ij}.
\end{aligned}$$

The scaling factors cancel out, meaning the scaled parameter $\hat{a}^*_{ij}$ is identical to the original a^*_{ij} in the Baum-Welch algorithm (5). This property applies to the observation probability matrix **B** and the initial probability **p** of the HMM. To derive the likelihood from the log-likelihood, simply take the exponential of the log-likelihood. The Baum-Welch functions for a continuous HMM are provided below, assuming a normal distribution for the observations, along with their supporting functions. The input parameters for the Baum-Welch algorithm are the initial parameters, and the output is the optimal set of parameters that maximize the likelihood of the observations.

```
# Baum-Welch function calibrates parameters for a continuous HMM
# assuming Gaussian observation probabilities.
# Input:
#   x       : observation sequence
#   A       : transition matrix
#   p       : initial state probabilities at time t = 1
#   mean    : vector of Gaussian means
#   sigma   : vector of Gaussian standard deviations
# Output:
#   A list containing updated HMM parameters
source("Cont.Estep.R")
source("Cont.Mstep.R")
source("scaled.forward.back.R")
Cont.BaumWelch <- function(x, A, p, mean, sigma) {
  maxiter <- 500          # maximum number of iterations
  tol <- 1e-05            # convergence tolerance
  prt <- TRUE             # print flag (unused here)
  posdiff <- TRUE         # enforce non-decreasing log-likelihood
  converge <- expression(diff < tol)
  m <- nrow(A)
  n <- length(x)
```

```
  oldLL <- -Inf          # initialize log-likelihood

  for (iter in 1:maxiter) {
    cond <- Cont.Estep(x, A, p, mean, sigma)
    diff <- cond$LL - oldLL
    if (diff < 0 && posdiff)
      stop("Worse log-likelihood on last iteration")
    if (eval(converge))
      break
    # Update transition matrix
    A <- diag(1 / apply(cond$v, MARGIN = 2, FUN = sum)) %*%
         apply(cond$v, MARGIN = c(2, 3), FUN = sum)
    # Update initial state probabilities
    p <- cond$u[1, ]
    # Update emission parameters
    ES <- Cont.MStep(x, cond)
    mean <- ES$mean
    sigma <- ES$sigma
    oldLL <- cond$LL
  }
  object <- list(A = A, p = p, mean = mean, sigma = sigma,
                 u = cond$u, v = cond$v, LL = cond$LL,
                 diff = diff, iter = iter)
  return(object)
}
```

Next, we present three functions—Cont.Estep, Cont.Mstep, and scaled.forward.backward which are utilized within the Cont.BaumWelch function. First, the scaled.forward.back function combines the forward and backward algorithms to calculate the likelihood of observations, as well as the logalpha and logbeta values of the HMM. These values are then used to calculate the conditional probability functions $\gamma_t(i)$ and $\xi_t(ij)$ within the Cont.Estep function.

```
# scaled.forward.back function calculates logalpha, logbeta,
# and log-likelihood for a continuous HMM with Gaussian emissions.
# Input:
#   x       : observation sequence (length n)
#   A       : transition matrix (m x m)
#   p       : initial state probabilities (length m)
#   mu      : vector of means for Gaussian emissions (length m)
#   sigma   : vector of standard deviations for Gaussian emissions
#             (length m)
# Output:
#   A list containing logalpha, logbeta, and total log-likelihood
scaled.forward.back <- function(x, A, p, mu, sigma) {
  m <- nrow(A)
  n <- length(x)
  # Compute emission probabilities
  prob <- matrix(0, nrow = n, ncol = m)
  for (k in 1:m)
    prob[, k] <- dnorm(x, mean = mu[k], sd = sigma[k], log = FALSE)
  # Forward pass (logalpha)
```

```
  phi <- as.double(p)
  logalpha <- matrix(0, nrow = n, ncol = m)
  llike <- 0
  for (i in 1:n) {
    if (i > 1)
      phi <- phi %*% A
    phi <- phi * prob[i, ]
    sumphi <- sum(phi)
    phi <- phi / sumphi
    llike <- llike + log(sumphi)
    logalpha[i, ] <- log(phi) + llike
  }
  LL <- llike
  # Backward pass (logbeta)
  logbeta <- matrix(0, nrow = n, ncol = m)
  phi <- rep(1 / m, m)
  llike <- log(m)
  for (i in seq(n - 1, 1, -1)) {
    phi <- A %*% (prob[i + 1, ] * phi)
    sumphi <- sum(phi)
    phi <- phi / sumphi
    llike <- llike + log(sumphi)
    logbeta[i, ] <- log(phi) + llike
  }
  return(list(logalpha = logalpha, logbeta = logbeta, LL = LL))
}
```

The following is the scaled Cont.Estep function that calculates the probability functions $\gamma_t(i)$ and $\xi_t(ij)$ defined in (2.18)and (2.20) respectively.

```
# Scaled Cont.Estep function calculates the conditional
# expectations and log-likelihood of a continuous HMM
# using scaled.forward.back function
# Input:
#   x      : observation vector
#   A      : transition matrix
#   p      : initial state probabilities
#   mean   : vector of Gaussian means
#   sigma  : vector of Gaussian standard deviations
# Output:
#   gamma  : posterior state probabilities
#   xi     : joint transition probabilities
#   LL     : log-likelihood of observations
# Call the scaled.forward.back function
source("scaled.forward.back.R")
Cont.Estep <- function(x, A, p, mean, sigma) {
  m <- nrow(A)
  n <- length(x)
  # Run scaled forward-backward algorithm
  h <- scaled.forward.back(x, A, p, mean, sigma)
  logalpha <- h$logalpha
  logbeta <- h$logbeta
```

```
  LL <- h$LL
  # Compute gamma (posterior state probabilities)
  gamma <- exp(logalpha + logbeta - LL)
  # Initialize xi (joint transition probabilities)
  xi <- array(NA, dim = c(n - 1, m, m))
  for (k in 1:m) {
    logprob <- dnorm(x[-1], mean[k], sigma[k], log = TRUE)
    logA <- matrix(log(A[, k]), byrow = TRUE, nrow = n - 1,
                   ncol = m)
    logPbeta <- matrix(logprob + logbeta[-1, k], byrow = FALSE,
                       nrow = n - 1, ncol = m)
    xi[, , k] <- logA + logalpha[-n, ] + logPbeta - LL
  }
  # Convert log-space xi to probability space
  xi <- exp(xi)
  return(list(gamma = gamma, xi = xi, LL = LL))
}
```

Last, the scaled Cont.MStep function re-estimates the HMM parameters in the Baum-Welch algorithm using the equations (2.28-2.29).

```
# Cont.MStep function updates the emission parameters of a
# continuous HMM, using the posterior state probabilities
# from the E-step function
# Input:
#   x    : observation vector
#   cond : object returned from Cont.Estep containing gamma (u)
# Output:
#   mean  : updated Gaussian means
#   sigma : updated Gaussian standard deviations
Cont.MStep <- function(x, cond) {
  m <- ncol(cond$u)
  n <- length(x)
  # Update means
  mean <- as.numeric(matrix(x, nrow = 1) %*% cond$u) /
          apply(cond$u, MARGIN = 2, FUN = sum)
  # Update standard deviations
  centered <- matrix(x, nrow = n, ncol = m) -
              matrix(mean, nrow = n, ncol = m, byrow = TRUE)
  weighted_var <- apply(centered^2 * cond$u, MARGIN = 2, FUN = sum)
                  /apply(cond$u, MARGIN = 2, FUN = sum)
  sigma <- sqrt(weighted_var)
  return(list(mean = mean, sigma = sigma))
}
```

We will use the Cont.BaumWelch functions to calibrate parameters for the continuous HMM presented in Example 2.5.

Example 2.5 (Find paymasters for an HMM) *The average temperatures in a city over one week are:*

$$O = \{68°\text{F}, 64°\text{F}, 61°\text{F}, 60°\text{F}, 57°\text{F}, 52°\text{F}, 48°\text{F}\}.$$

Suppose that the temperatures follow a continuous HMM with three hidden states, find the parameters of the model.

Solution 2.5

The Cont.BaumWelch function can solve the example instantly. First, we use the following Initial function to calculate the initial parameters of the model. Readers can change the initial parameters in the Initial function as long as they satisfy the conditions in (1.19).

```
# Initial function estimates starting parameters for a continuous
# HMM based on input data and number of hidden states.
# Inputs:
#   y : observation vector
#   N : number of hidden states
# Output:
#   A list containing initial estimates for A, p, mean, and sigma
Initial <- function(y, N) {
  m <- mean(y)
  s <- sd(y)
  mean <- rep(m, N) + rnorm(N) * s
  p <- c(1, rep(0, N - 1))
  sigma <- rep(s, N)
  A <- matrix(1 / N, nrow = N, ncol = N)
  IV <- list(A = A, p = p, mean = mean, sigma = sigma)
  return(IV)
}
```

Then, we apply the scaled Baum-Welch algorithm to determine the optimal parameters for the model.

```
# Calibrate HMM parameters using the Baum-Welch algorithm
# Load required functions
source("C:/Users/scaled.forward.back.R")
source("C:/Users/Cont.Estep.R")
source("C:/Users/Cont.Mstep.R")
source("C:/Users/Cont.BaumWelch.R")
source("C:/Users/Initial.R")
# Input observation sequence (weekly temperatures in Fahrenheit)
x <- c(68, 64, 61, 60, 57, 52, 48)
# Initialize parameters for a continuous HMM with 3 states
Par <- Initial(x, 3)
# Run Baum-Welch algorithm to calibrate HMM parameters
l <- Cont.BaumWelch(x, Par$A, Par$p, Par$mean, Par$sigma)
# Print the results
print(l)
```

We got the output which includes the HMM parameters, the expectation probability functions $\gamma_t(i)$ 2.19 presented by (u_i), and $\xi_t(ij)$ 2.21 presented by (v_{ij}), the log likelihood of the observations, the error of the estimations (diff), and the number of interactions (iter) in Example 2.5.

```
  $A
              [,1]          [,2]          [,3]
[1,] 5.051563e-01 4.948437e-01 2.095076e-20
[2,] 5.186625e-46 4.788061e-01 5.211939e-01
[3,] 4.628333e-85 3.329005e-44 1.000000e+00

$p
[1] 1 0 0

$mean
[1] 65.94844 60.51033 52.48486

$sigma
[1] 2.0527799 0.4998932 3.7982074
$u (gamma)
              [,1]          [,2]          [,3]
[1,] 1.000000e+00  0.000000e+00 0.000000e+00
[2,] 1.000000e+00  1.568998e-10 1.424842e-24
[3,] 2.087615e-02  9.791238e-01 3.456333e-13
[4,] 1.960695e-14  9.399260e-01 6.007402e-02
[5,] 7.472639e-28  1.358864e-10 1.000000e+00
[6,] 1.491715e-34  8.240577e-71 1.000000e+00
[7,] 1.491052e-34 2.393378e-153 1.000000e+00

$v (xi)
, , 1

              [,1]          [,2]          [,3]
[1,] 1.000000e+00  0.000000e+00  0.000000e+00
[2,] 2.087615e-02  3.541972e-57 6.791025e-108
[3,] 1.960695e-14  9.159379e-58 2.450947e-108
[4,] 7.472639e-28  5.494711e-69 1.633341e-109
[5,] 1.491715e-34  1.119560e-81 3.831868e-111
[6,] 1.491052e-34 7.643158e-132 4.313743e-101

, , 2

               [,1]          [,2]          [,3]
[1,]  1.568998e-10  0.000000e+00  0.000000e+00
[2,]  9.791238e-01  1.565542e-10  2.338676e-65
[3,]  2.087615e-02  9.190498e-01  1.916117e-55
[4,]  1.960695e-14  1.358668e-10  3.146737e-55
[5,]  8.182702e-71  5.787496e-73 1.543366e-106
[6,] 2.393378e-153 1.156176e-205 5.084174e-179

, , 3

              [,1]          [,2]          [,3]
[1,] 1.424842e-24 0.000000e+00 0.000000e+00
[2,] 8.407763e-23 3.456333e-13 1.424842e-24
```

```
[3 ,] 5.307501e-23 6.007402e-02 3.456333e-13
[4 ,] 5.275727e-24 9.399260e-01 6.007402e-02
[5 ,] 7.472638e-28 1.358864e-10 1.000000e+00
[6 ,] 6.634940e-38 8.240577e-71 1.000000e+00


$LL
[1] -16.54047

$diff
[1] 8.661128e-06

$iter
[1] 13
```

With the maximum number of interactions of 500 (maxiter=500), and the tolerance of 10^{-5}, (tol=1e-05), the codes found the HMM parameters after only 13 iterations (iter=13).

In Example 2.5, the observation temperatures are picked randomly without knowing their distributions. Thus, we have no way to evaluate the HMM parameters. We will present another example where we know the distributions of the observations. Therefore, we can compare the calibrated parameters with the actual ones.

Example 2.6 *Given fifty observations randomly generated from one of two Gaussian distributions, the means μ and standard deviations σ are:*

$$\mu = (\mu_1, \mu_2) = (-1, 2) \tag{2.79}$$

$$\sigma = (\sigma_1, \sigma_2) = (0.3, 1.2) \tag{2.80}$$

Suppose that you don't know the two normal distributions, and you only know that the observations come from a continuous HMM with two stats. Calibrate the HMM parameters.

Solution 2.6
We will use the Cont.BaumWelch function to solve the problem.

```
# Run functions used in the Baum-Welch algorithm
source("C:/Users/scaled.forward.back.R")
source("C:/Users/Cont.Estep.R")
source("C:/Users/Cont.Mstep.R")
source("C:/Users/Cont.BaumWelch.R")
source("C:/Users/Initial.R")
# Generate observation data from two normal distributions
mu1 <- c(-1, 2)
sigma1 <- c(0.3, 1.2)
N <- 50
y <- floor(runif(N) * 2 + 1)  # Hidden states: 1 or 2
x <- numeric(N)

for (i in 1:N) {
  if (y[i] == 1)
    x[i] <- rnorm(1, mean = mu1[1], sd = sigma1[1])
  else
    x[i] <- rnorm(1, mean = mu1[2], sd = sigma1[2])
```

```
}
# Calculate initial parameters for the continuous HMM with
# 2 states
Par <- Initial(x, 2)
# Print the initial parameters
print(Par)
# Calibrate HMM parameters using Baum-Welch
l <- Cont.BaumWelch(x, Par$A, Par$p, Par$mean, Par$sigma)
# Print the results
print(l)
```

The output of the code are:

```
    > Par: The initial parameters
$A
      [,1] [,2]
[1,]  0.5  0.5
[2,]  0.5  0.5

$p
[1] 1 0

$mean
[1] 1.665767 1.149696

$sigma
[1] 1.892973 1.892973
The caliberated paramester by using the baumWelch1 algorithm
> l$A
          [,1]      [,2]
[1,] 0.4625533 0.5374467
[2,] 0.8136014 0.1863986
> l$p
[1] 1 0
> l$mean
[1] -1.098135 2.006722
> l$sigma
[1] 0.3489578 1.3846088
```

The results show that, starting with the initial parameters, the scaled Baum-Welch algorithm identified new parameters where the means and standard deviations of the two Gaussian distributions closely match the actual distributions used to generate the observation sequence in (2.79) and (2.80). Specifically, the original distributions from which the observations were generated are $\mathcal{N}(-1, 0.3)$ and $\mathcal{N}(2, 1.2)$, while the calibrated distributions have means and standard deviations of $\mathcal{N}(-1.098, 0.349)$ and $\mathcal{N}(2.007, 1.385)$, respectively. Figure 2.5 is the histogram of fifty thousand observations generated from two normal distributions with means and standard deviations listed in equation (2.79) and equation (2.80). You can see clearly the two distributions on the figure. With the observations, the scaled Baum-Welch algorithm found the following HMM parameters:

```
> l$A
          [,1]      [,2]
[1,] 0.4288207 0.5711793
[2,] 0.6531681 0.3468319
> l$p
[1] 1 0
> l$mean
[1] −1.066727 1.833881
> l$sigma
[1] 0.2549729 0.9769185
```

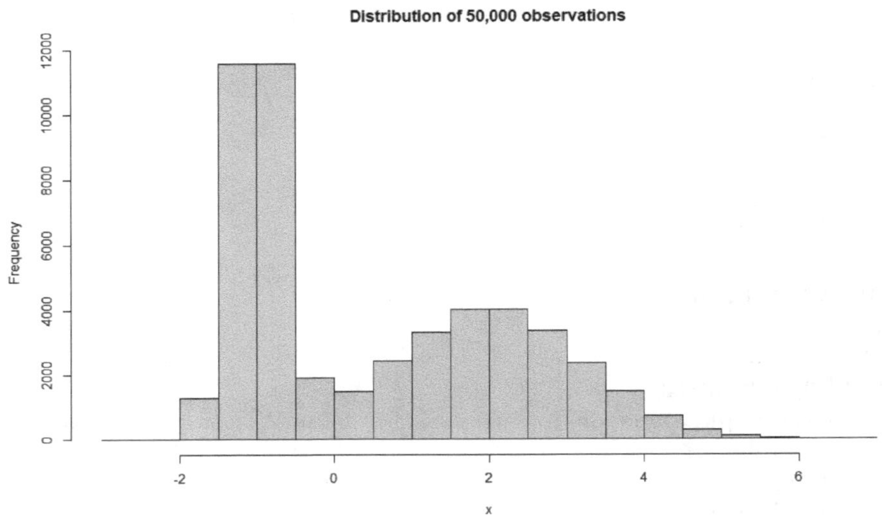

FIGURE 2.5
Histogram of the 50,000 observations randomly generated from two normal distributions with means and standard deviations in (2.79) and (2.80).

3

Model Selection

From the definition of the HMM, we know that an HMM consists of two important factors: the number of hidden states and the emission distributions of observations. In this chapter we will present some methods to choose the number of states for an HMM, to detect the emission distributions for the given data, and to choose the sample size of the data. We will then present a formula to compare two HMMs. Finally, after defining the HMM parameters, we will present some evaluation indicators to evaluate the chosen HMM.

3.1 Model Selection Indicators

Choosing a number of hidden states for an HMM is a critical task. We use four common model criteria, the Akaike information criterion (AIC), the Bayesian information criterion (BIC), the Hannan-Quinn criterion (HQC), and the Consistent AIC (CAIC) to evaluate the performances of the HMM with different numbers of states. The AIC was introduced by the Japanese statistician Akaike in [2]. The criteria measures how well a model fits with a given data set by evaluating the log likelihood of the model. Among a collection of models for the data set, the model that has the lowest AIC is the best model. BIC was proposed as a criterion for model selection among a finite set of models [124]. Similar to the AIC, the model with a lower BIC is better. The HQC was presented in [60]. The authors added more weight to the penalty term in the HQC compare to the one in the AIC. The CAIC, which is the modifier of the BIC, was introduced in [19]. In all of these four criteria, a model with lower criterion value is better. The four criteria are suitable for the HMM because in the model training algorithm and the Baum-Welch algorithm, the expectation–maximization (EM) method was used to maximize the log likelihood, $\ln(L)$, of the model. Recall that the likelihood, L, of the HMM is the probability of observation sequence O given the model's parameters λ

$$L = \mathbb{P}(O|\lambda).$$

These criteria AIC, BIC, HQC, and CAIC are calculated using the following formulas respectively:

$$\text{AIC} = -2\ln(L) + 2k. \tag{3.1}$$

$$\text{BIC} = -2\ln(L) + k\ln(M). \tag{3.2}$$

$$\text{HQC} = -2\ln(L) + 2k\ln(\ln(M)). \tag{3.3}$$

$$\text{CAIC} = -2\ln(L) + k(\ln(M) + 1), \tag{3.4}$$

where L is the likelihood of the model, n is the number of observation points, and k is the number of estimated parameters in the model. From the formulas, we see that AIC doesn't count the observation size, while the BIC does. In the HMM, we need to choose number of hidden states, which will determine the number of parameters of the model, and the size of observation data. If we only want to choose number of states for the HMM, the AIC is the appropriate criteria. If we want to determine the sample size then the BIC should be used. Schwarz wrote on his paper [124] that: "Quantitatively, since our procedure differs from Akaike's only in that the dimension is multiplied by $\frac{1}{2}\log n$, our procedure leans more than Akaike's towards lower-dimensional models (when there are 8 or more observations). For large numbers of observations the procedures differ markedly from each other." Tu and Xu in [137] implemented the four criteria and ranked them in order from weak to strong: AIC, HQC, BIC and CAIC. We will use all of the criteria to choose number of states for an HMM.

If the distribution of emission probability of an observation corresponds to its hidden state is a Gaussian, the number of parameters, k, is formulated as $N^2 + 2N - 1$, where N is the number of states used in the HMM. In the AIC if more parameters are added to the model, the second term is bigger and the log likelihood of the model increases, which makes the first term smaller. However, when the sample size is large, the log likelihood dominates the penalty term, $2k$. Thus, the AIC will prefer the model with more parameters. In the BIC, the penalty term (the second term) includes the log of sample size and number of parameters. Thus, selecting the "best fit" model using BIC is equivalent to choosing the model with the largest log likelihood and appropriate sample size. Note that when using the BIC, we may need to keep k or n fixed when letting the remainder variable float in order to choose its optimal value. For example, you can keep k the same to find the optimal n^*. After that, keep n^* the same to final the optimal k^*.

3.2 Choosing the Number of States for an HMM

In this example, we choose one of the common benchmarks for the U.S. stock market, the Standard & Poor's 500 index (S&P 500) to be our observation data. The S&P 500 was monthly data from January 2006 to November 2016 and was taken from `finance.yahoo.com`. We assume that the stock prices follow a continuous HMM with a Gaussian distribution for an observation corresponding to its state. We will use the four criteria: the AIC, the BIC, the HQC, and the CAIC to choose a number of states for our HMM. We limit numbers of states from two to six to keep the model simple and feasible for stock prediction. Below are the plot criteria for the SPY index with 120 calibrations of the model's parameters.

Look at the Figures 3.1 and 3.2, we can see that the CAIC gave the most consistent relationships between the five HMMs. Following is the BIC while the AIC and the HQC gave the less consistent results. Using the CAIC we may choose the HMM with five states because its values are the lowest among the five models for most of the 120 calibrations. Using the BIC criteria we may choose the HMM with the four states. Using the AIC and the HQC, it would be hard to choose between an HMM with four or five states. The results can only be applied for data used in the test. The chosen models may not work well with other data sets. Thus, for each data set, we should use one of the criteria to choose the best HMM. In the event we are undecided between two models, we can use method in the next section to compare the HMMs.

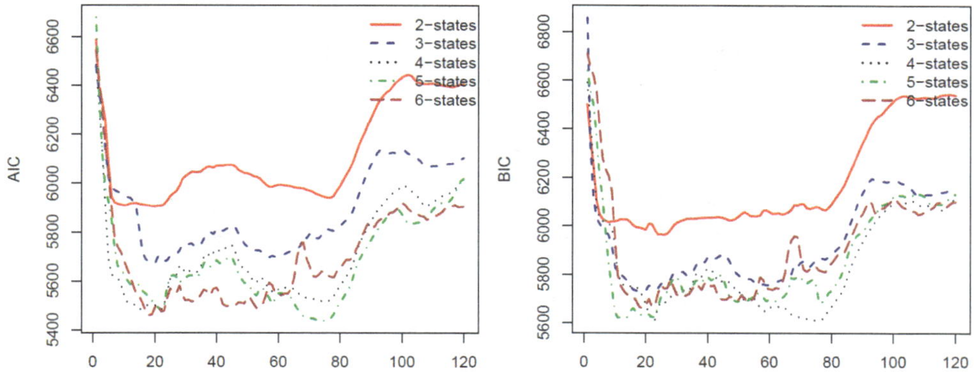

FIGURE 3.1
AIC (left) and BIC (right) for 120 HMM parameter calibrations using SPY monthly prices.

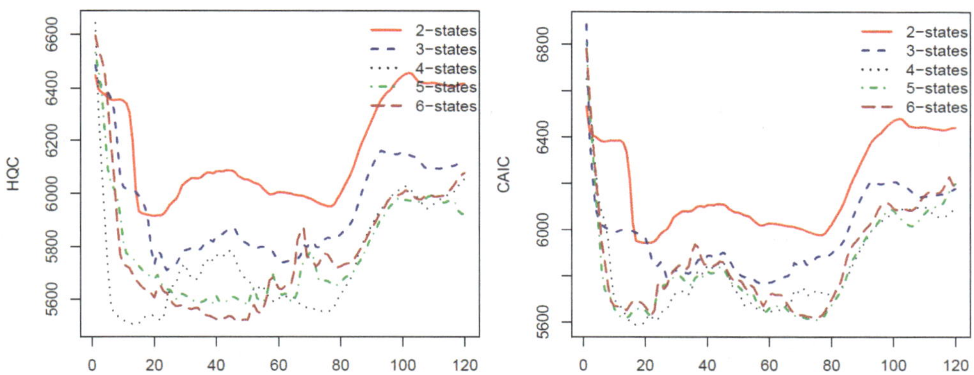

FIGURE 3.2
HQC (left) and CAIC (right) for 120 HMM parameter calibrations using S&P 500 monthly prices.

3.3 Comparison of HMMs

Rabiner [115] presented a formula to compare two HMMs. Suppose that we have the two HMMs with parameters λ_1 and λ_2. Rabiner first defined the distance $D(\lambda_1, \lambda_2)$ by the formula:

$$D(\lambda_1, \lambda_2) = \frac{\mathbf{1}}{\mathbf{T}}\left[\log\left(\mathbb{P}\left(\mathbf{O}^{(2)}|\lambda_1\right)\right) - \log\left(\mathbb{P}\left(\mathbf{O}^{(2)}|\lambda_2\right)\right)\right], \tag{3.5}$$

where $O^{(2)} = \{O_1, O_2, ..., O_T\}$ is the sequence of observations generated by the model with parameter λ_2 by following the process 1.3.3 from Chapter 1. The distance $D(\lambda_1, \lambda_2)$ evaluates how well the model λ_1 matches observations generated by model λ_2. Similarly, the distance

$$D(\lambda_2, \lambda_1) = \frac{1}{T}\left[\log\left(\mathbb{P}\left(O^{(1)}|\lambda_1\right)\right) - \log\left(\mathbb{P}\left(O^{(1)}|\lambda_2\right)\right)\right], \tag{3.6}$$

where $O^{(1)} = \{O_1, O_2, ..., O_T\}$ is the sequence of observations generated by the model with parameter λ_1. The distance measures how well the model λ_2 matches observations generated by the model λ_1. Then, the distance between the two models was defined by the symmetry formula:

$$D_s(\lambda_1, \lambda_2) = \frac{D(\lambda_1, \lambda_2) + D(\lambda_2, \lambda_1)}{2}. \tag{3.7}$$

In Ref. [115], the author also gave an example of how two HMMs with different parameters λ_1 and λ_2 can be statistically equivalent. Consider the two 2-state models

$$\lambda_1 = \{\mathbf{A_1}, \mathbf{B_1}, \mathbf{p_1}\}, \lambda_2 = \{\mathbf{A_2}, \mathbf{B_2}, \mathbf{p_2}\},$$

with

$$\mathbf{A_1} = \begin{bmatrix} a & 1-a \\ 1-a & a \end{bmatrix}, \mathbf{B_1} = \begin{bmatrix} b & 1-b \\ 1-b & b \end{bmatrix}, \mathbf{p_1} = [.5, .5],$$

and

$$\mathbf{A_2} = \begin{bmatrix} c & 1-c \\ 1-c & c \end{bmatrix}, \mathbf{B_2} = \begin{bmatrix} d & 1-d \\ 1-d & d \end{bmatrix}, \mathbf{p_2} = [\mathbf{0.5}, \mathbf{0.5}],$$

where

$$0 \leq a, b, c, d \leq 1. \tag{3.8}$$

If the two models are statistically equivalent, for any observation O_t we must have the probability of the observations that have symbol v_k on each model are equal. That means,

$$\begin{aligned} \mathbb{E}[O_t = v_k|\lambda_1] &= \mathbb{E}[O_t = v_k|\lambda_2], \text{for any } \mathbf{v_k}, \text{or} && (3.9) \\ ab + (1-a)(1-b) &= cd(1-c)(1-d) \text{ or} && (3.10) \\ d &= \frac{a+b-2ab}{1-2c}. && (3.11) \end{aligned}$$

We can pick a set of a, b, c, d that satisfies conditions (3.8), and (3.11), and we will have two statistical equivalent models. For example, if $a = 0.6, b = 0.5, c = 0.1$, then

$$d = \frac{0.6 + 0.5 - 2(0.6)(0.5)}{1 - 2(0.1)} = \frac{0.5}{0.8} = 0.625.$$

Thus, we have two statistical equivalent models:

$$\lambda_1 = \left\{ \begin{bmatrix} 0.6 & 0.4 \\ 0.4 & 0.6 \end{bmatrix}, \begin{bmatrix} 0.5 & 0.5 \\ 0.5 & 0.5 \end{bmatrix}, [0.5, 0.5] \right\}, \tag{3.12}$$

and

$$\lambda_2 = \left\{ \begin{bmatrix} 0.1 & 0.9 \\ 0.9 & 0.1 \end{bmatrix}, \begin{bmatrix} 0.625 & 0.373 \\ 0.375 & 0.625 \end{bmatrix}, [0.5, 0.5] \right\}. \tag{3.13}$$

3.4 Model Evaluation Indicators

HMMs can be used to predict future outcomes for events in other areas such as economics or weather forecast. In this section, we will present three common model evaluation indicators to evaluate predictive models for numerical data predictions: the mean absolute percentage error (MAPE), the average absolute error (AAE), and the root-mean-square error (RMSE). These error estimators are calculated using the following expressions:

$$\text{MAPE} = \frac{1}{N}\sum_{t=1}^{N}\left|\frac{A_t - F_t}{A_t}\right|, \tag{3.14}$$

$$\text{AAE} = \frac{1}{N}\sum_{t=1}^{N}\left|A_t - F_t\right|, \tag{3.15}$$

$$\text{RMSE} = \sqrt{\frac{1}{N}\sum_{t=1}^{N}(A_t - F_t)^2}, \tag{3.16}$$

where N is number of observation points, A_t is the actual observation value at time t, and F_t is the forecast of observation value at time t. We can see from the formulas of the error measurements that the MAPE is actually the mean of the absolute relative error. Therefore, its value ranges from zero to relative one. The AAE and RMSE measures the average difference between predicted values and the actual values. The range of the two values is from zero to infinity. The AAE and RMSE are more sensitive to outliers than the MAPE. For all three error estimators, a lower value indicates a better predictive model.

In the next chapter, we will apply the model selection methods to find the best HMM for solving real-life problems for both single and multiple observation sequences.

4

Applications of HMM in Finance

In this chapter, we provide a brief overview of the applications of hidden Markov models (HMMs) in finance, with a focus on their use in stock price prediction, mortgage-backed securities, and stock trading. Additionally, we will highlight key differences between Markov chains, regime-switching models, and HMMs - distinctions that have often been overlooked in some published works on the HMMs and their applications in finance.

4.1 Overview of Published Work on HMM Applications in Finance

In this section, we will give an overview of published articles and books about the HMM and its applications in finance.

4.1.1 Published Articles on HMM Applications in Finance

Recently, researchers have applied the HMM to predict stock prices. The HMM was used to forecast stock price for interrelated markets in [61]. The HMM with two states was used to predict regimes in market turbulence, inflation and industrial production index in [79]. Nguyen used the HMM with both single and multiple observations to forecast economic regimes and stock prices in [103]. The following year, the HMM was used for single observation data to predict regimes of some economic indicators and stock selections were made based on the performances of these stocks during the predicted regimes in [105]. The HMM was also used to analyze stock market trends in [78]. Previously, the HMM was applied to predict S&P daily prices to trade stock in [82]. The HMM algorithms were used for forecasting cross currency in [67]. However, the author indicated that the strategy they developed was "too unstable" to be considered as a trading tool for foreign exchange data.

4.1.2 Books on HMM and Its Applications in Finance

Aggoun, Elliott, and Moore wrote the book titled: "Hidden Markov Models Estimation and Control" in 1995 [40]. The authors first gave optimal estimation algorithms of the HMM for the discrete-time, discrete-state cases, then focused on the optimal control algorithms of the model for continuous-range states and continuous-time HMM models. The book heavily presented the HMM algorithms in theoretical probability principles. Recently, in 2016, Zucchini, MacDonald, and Langrock published the book titled: "Hidden Markov Models for Time Series, An Introduction Using R", Second Edition [165]. In this book the authors presented the basic model formulation and its applications to many areas, and had some R code that were applied for some data sets. However, the model's applications in the book are mainly focused on decoding and predicting the probability of upcoming states. It does not provide a technique of using the HMM to predict future observations. The book titled:

"Hidden Markov Model in Finance" was presented in [97] in 2007. However, the book is a collection of papers that used the regime-switching model introduced in [59], which is not identical to the HMM introduced in [12]. None of the papers actually use the HMM that we have discussed in this publication. Therefore, the title of the book will lead to a misunderstanding of the HMM. By the definitions of the HMM and the Markov chain in Chapter 1, we know that the HMM consists of a hidden Markov chain process and a stochastic observation process. Thus, the Markov chain process is only the hidden state sequence of the HMM. So, we should not consider a model that uses Markov chain as an HMM if the observation sequence does not satisfy the conditions of the HMM. We will present some main concepts of the Regime Switching Model (RSM) and the HMM in the next section to highlight the differences between the two models. We will present the key concepts and its applications of the RSM in the following sections to compare it with the HMM.

4.1.2.1 Regime Switching Model vs. HMMs

Based on the principal concepts of state sequences, Hamilton [59] developed a RSM for nonstationary time series and the business cycle. In [59], observation data was assumed to be associated with a first-order Markov process with two states, or regimes. However, the observation corresponding with each regime at time t depends on the current state and previous observation points. In contrast, in the HMM in [12], an observation associated with a state depends on only the current state, which is defined by an emission probability 1.3.1 and 1.21. Although the RSM and HMM are both associated with regimes or hidden states, an observation at time t of the RSM should be viewed as a regression model of observations in previous time and regime-shift variables. Therefore, the HMM and RSM are different in their structures of observation sequences and the hidden states. The differences of the two models were mentioned in [95].

4.1.2.2 Applications of Regime-Switching Models

The RSM is commonly used for time series related problems. Investigators have been using the RSM to detect financial crises for decades. Bonnie [18] investigated the dynamic impact of macroeconomic aggregates on housing prices and housing stocks. Elliott and Wilson [41] used the RSM to model short-term interest rates, which assumed that the mean-reverting level follows a finite-state and continuous-time Markov chain. Garcia and Parron [53] had used a three-state RSM and a regression method to estimate the U.S. real interest rate and inflation rate. In the early 2000s, Ang and Bekaert [5] applied a RSM for international asset allocation. Guidolin and Zimmerman [57] used a four-state RSM to study asset allocation decisions on asset returns. In 2007, Erlwein and Mamon [44] implemented the HMM for a financial data set of 30-day Canadian Treasury bill yields. More recently, in 2013, Zhu and Cheng [164] used the RSM to examine macroeconomic risk level. That same year, Nneji, Brooks and Ward [107] used the RSM for investigating the impact of the macro economy on the dynamics of the residential real estate market in the U.S.

4.2 R Functions for HMM Implementation

In this section, we present all the R functions used in this chapter. These functions are designed to work with real-market continuous data, involving single or multiple observation sequences. Specifically, we will cover the HMM functions: forward, backward, Viterbi, Baum-Welch, and Initial, tailored for continuous data with multiple observation sequences. The algorithms underlying these functions were introduced in Chapter 2.

The first function calculates the initial parameters of the HMM based on the input data. This function is identical to the Initial function introduced in Chapter 2.

```
# Initial function generates initial parameters for a continuous
# HMM.
# Input:
#   y : observation data (numeric vector)
#   N : number of hidden states
# Output:
#   IV: list containing initialized parameters (A, p, mean, sigma)
Initial <- function(y, N) {
  m <- mean(y)
  s <- sd(y)
  b <- N - 1
  # Initialize means with random perturbation
  mean <- rep(m, N) + rnorm(N) * s
  # Initial state probabilities (first state only)
  p <- c(1, rep(0, b))
  # Standard deviations
  sigma <- rep(s, N)
  # Transition matrix with uniform probabilities
  A <- matrix(1 / N, nrow = N, ncol = N)
  # Return as a named list
  IV <- list(A = A, p = p, mean = mean, sigma = sigma)
  return(IV)
}
```

Next is the Estep function, which computes the conditional expectation probabilities γ and ξ, the transition matrix, and the log-likelihood of the observation probabilities for a continuous HMM with N states. This version differs slightly from the Cont.Estep function presented in Chapter 2, as it also returns the updated transition matrix of the HMM.

```
# Estep.Ndata function calculates conditional expectations and
# log-likelihood for a continuous HMM using scaled
# forward-backward algorithm in log-space.
# Input:
#   x       : observation sequence (length n)
#   A       : transition matrix (m x m)
#   p       : initial state probabilities (length m)
#   mean    : vector of Gaussian means (length m)
#   sigma   : vector of Gaussian standard deviations (length m)
#   N       : number of hidden states (used for initialization)
# Output:
#   S  : updated transition count matrix (m x m)
#   u  : posterior state probabilities (gamma)
#   v  : joint transition probabilities (xi)
#   LL : log-likelihood of the observation sequence
# Load all required functions before running Estep.Ndata
# Load the scaled.forward.backward function
source("C:/Users/scaled.forward.back.R")
# If scaled.forward.back.R uses helper functions, load them too
Estep.Ndata <- function(x, A, p, mean, sigma, N) {
```

```
  m <- nrow(A)
  n <- length(x)
  # Initialize transition count matrix
  S <- matrix(0, nrow = m, ncol = m)
  # Run scaled forward-backward algorithm
  h <- scaled.forward.back(x, A, p, mean, sigma)
  logalpha <- h$logalpha
  logbeta <- h$logbeta
  LL <- h$LL
  # Compute gamma (posterior state probabilities)
  u <- exp(logalpha + logbeta - LL)
  # Compute xi (joint transition probabilities)
  v <- array(NA, dim = c(n - 1, m, m))
  for (k in 1:m) {
    logprob <- dnorm(x[-1], mean[k], sigma[k], log = TRUE)
    logA <- matrix(log(A[, k]), nrow = n - 1, ncol = m,
                   byrow = TRUE)
    logPbeta <- matrix(logprob + logbeta[-1, k],
                   nrow = n - 1, ncol = m, byrow = FALSE)
    v[, , k] <- logA + logalpha[-n, ] + logPbeta - LL
  }
  v <- exp(v)
  S <- apply(v, MARGIN = c(2, 3), FUN = sum)
  return(list(S = S, u = u, v = v, LL = LL))
}
```

The Estep.Ndata function will be used in the following function ESTEP.N.DATA to calculate the HMM parameters for multiple observation sequences.

```
# ESTEP.N.DATA function estimates HMM parameters from multiple
# observation sequences.
# Input:
#   y      : 3D array of observation sequences (K x T x 1)
#   T      : length of each observation sequence
#   K      : number of sequences
#   N      : number of hidden states
#   A      : transition matrix (N x N)
#   p      : initial state probabilities (length N)
#   mean   : vector of Gaussian means (length N)
#   sigma  : vector of Gaussian standard deviations (length N)
# Output:
#   mean   : updated means
#   sigma  : updated standard deviations
#   S      : updated transition matrix
#   p      : updated initial probabilities
#   LL1    : total log-likelihood
#   u1     : averaged posterior state probabilities
# Load the Estep.Ndata function
source("C:/Users/Estep.Ndata.R")
# Load Estep.Ndata.R helper functions
source("C:/Users/scaled.forward.back.R")
ESTEP.N.DATA <- function(y, A, p, mean, sigma, T, K, N) {
```

```
  m <- nrow(A)
  S1 <- matrix(0, nrow = m, ncol = m)
  u1 <- matrix(0, nrow = T, ncol = m)
  LL1 <- 0
  mean.new <- matrix(0, nrow = 1, ncol = m)
  sigma.new <- matrix(0, nrow = 1, ncol = m)
  S2 <- matrix(0, nrow = m, ncol = m)
  for (k in 1:K) {
    x <- y[k, , ]
    UV <- Estep.Ndata(x, A, p, mean, sigma, N)
    u1 <- u1 + UV$u
    S1 <- S1 + UV$S
    LL1 <- LL1 + UV$LL
    S2 <- S2 + diag(1 / apply(UV$v, MARGIN = 2, FUN = sum))
    mean.new <- mean.new + as.numeric(matrix(x, nrow = 1) %*% UV$u)
    sigma.new <- sigma.new + apply(
      (matrix(x, nrow = T, ncol = m) -
        matrix(mean, nrow = T, ncol = m, byrow = TRUE))^2 * UV$u,
      MARGIN = 2, FUN = sum
    )
  }
  S <- S2 %*% S1
  S <- S / apply(S, 1, sum)
  mean <- mean.new / apply(u1, 2, sum)
  sigma <- sqrt(sigma.new / apply(u1, 2, sum))
  u1 <- u1 / K
  p <- u1[1, ]
  return(list(mean = mean, sigma = sigma, LL1 = LL1, S = S, p = p,
          u1 = u1))
}
```

Next is a simple function that finds the index of the minimum value in a vector, which will be used in other R functions.

```
# argmin function returns the index of the minimum value in a
# numeric vector
argmin <- function(v) {
  which.min(v)
}
```

Lastly, the BW.Ndata function calibrates the HMM parameters for multiple observation sequences.

```
# BW.Ndata function calibrates HMM parameters using the Baum-Welch
# algorithm for multiple observation sequences.
# Input:
#    y      : 3D array of observations (K x T x 1)
#    T      : length of each observation sequence
#    K      : number of sequences
#    N      : number of hidden states
#    A      : initial transition matrix (N x N)
#    p      : initial state probabilities (length N)
```

```
#   mean   : initial Gaussian means (length N)
#   sigma  : initial Gaussian standard deviations (length N)
# Output:
#   mean   : calibrated means
#   sigma  : calibrated standard deviations
#   S      : calibrated transition matrix
#   p      : calibrated initial probabilities
#   LL     : final log-likelihood
#   u      : final expected state probabilities
#   diff   : final change in log-likelihood
#   iter   : number of iterations performed
# Load the ESTEP.N.DATA function
source("C:/Users/ESTEP.N.DATA.R")
# Load ESTEP.N.DATA.R helper functions
source("C:/Users/Estep.Ndata.R")
# Load Estep.Ndata.R helper functions
source("C:/Users/scaled.forward.back.R")
BW.Ndata <- function(y, A, p, mean, sigma, T, K, N) {
  maxiter <- 500
  tol <- 1e-6
  oldLL <- -100
  diff <- 100
  iter <- 0
  while ((iter < maxiter) && (diff > tol)) {
    Var <- ESTEP.N.DATA(y, A, p, mean, sigma, T, K, N)

    diff <- Var$LL1 - oldLL
    A <- Var$S
    p <- Var$p
    mean <- Var$mean
    sigma <- Var$sigma
    oldLL <- Var$LL1
    u <- Var$u1
    iter <- iter + 1
  }
  object <- list(
    u = u,
    A = A,
    p = p,
    mean = mean,
    sigma = sigma,
    LL = Var$LL1,
    diff = diff,
    iter = iter
  )
  return(object)
}
```

4.3 HMM for Mortgage-Backed Security ETFs

In this section, we give a brief overview of mortgage-backed securities (MBS) and apply the HMM to analyze mortgage-backed securities exchange-traded funds (MBS ETFs). In the HMM application portion, we first use the Baum-Welch algorithms to calibrate the model's parameters and the Viterbi algorithm to locate the hidden states (or regimes) of observation data. We then describe a method of using the HMM to predict the probability of recession states and develop a strategy to predict some MBS ETFs. Finally, we use the HMM to select MBS ETF portfolios.

4.3.1 Overview of Mortgage-Backed Securities

MBS have been used as investment instruments for investors and lenders since the 1980s. The MBS industry provides banks with more cash to make more mortgage loans, keeps mortgage rates competitive, and keeps mortgages readily available. Many exchange-traded funds that focus exclusively on mortgage securities, called MBS ETFs, were formed to make mortgage investments more feasible for individual investors.

The U.S. housing market reached its peak in mid-2006, where it then began to decline. According to a 2007 report in [36], from 1997 to 2006, the price of a typical American house increased by 124 percent. The U.S. housing market experienced the subprime mortgage crisis during the economic recession of December 2007 to June 2009. After seven years of declining rates, from 6.825 percent in June 2006 to 3.682 percent in May 2013, the 30-year fixed-rate mortgage increased to 4.004 percent in August 2015. The changes in these interest rates have an inverse affect on MBS. If interest rates increase, MBS prices will decrease. Conversely, when interest rates drop, MBS securities will rise.

MBS and mortgage rates are driven by many economic factors, such as gross domestic product, inflation rates, and interest rates. Researchers have investigated the performance of some economic indicators to evaluate the MBS. Zenios [162] considered both the future interest rate and the prepayment activity of the MBS to simulate an MBS portfolio. Calhoun and Deng [22] analyzed the results of using different loan-level statistical models for fixed and adjustable rate mortgages. Kau, Keenan, Muller and Epperson [77] investigated many intricacies of a fixed-rate mortgage model and concluded that "default always lowers the value of the mortgage to the lender, whereas nonfinancial prepayment always raises it." In the following sections, we select some MBS ETFs and use the HMM to investigate their behaviors.

4.3.2 Data Selection

As mentioned in Section 4.3.1, MBS have been known to investors since the 1980s. However, individual investors had limited access to MBS until the mortgage crisis in 2007. Several ETFs that focus exclusively on MBS were provided for trading by many brokers. In this research, we collect historical data of MBS ETFs that are actively traded on the market, which can be downloaded from Yahoo Finance (https://finance.yahoo.com). These MBS ETFs were listed in Table 4.1. The data were gathered in three frequencies: daily, weekly and monthly. For each of these ETFs we have four observation sequences: "close," "open," "low" and "high" prices. We assume throughout this paper that the four observation sequences are independent. You can see from Table 4.1 that most of the MBS ETFs were inserted recently. Therefore, to show applications of the HMM to MBS ETFs in the following Sections 4.3.3-4.3.5, we will use the iShare MBS ETF known as MBB, because this ETF has long enough historical data to be useful.

TABLE 4.1
List of MBS ETFs

Broker	Trading Symbol	Issued Date
iShares	MBB	03/13/2007
	GNMA	02/14/2012
	CMB	02/14/2012
Vanguard	VMBS	11/23/2009
Barclays	MBG	01/30/2009
FlexShares	MBSD	01/04/2016
First Trust	LMBS	11/05/2014

4.3.3 Using an HMM to Identify Hidden States

In this section, we present the results of using an HMM to find hidden states (or regimes) of fund MBB. The observation price of MBB at time t is a continuous random variable. We assume that the observation probability $b_i(o_t) = \mathbb{P}(q_t = S_i, O_t|\lambda)$ is the density of the normal distribution with mean μ_i and standard deviation σ_i at point $O - t, 1 \leq i \leq N$. Therefore, the symbol ν_k is omitted in the continuous HMM, and the parameter of the model is $\lambda = \{\mathbf{A}, \mu, \sigma, \mathbf{p}\}$. To use the HMM, we first need to choose observation data and the number of states, N. To make it simple, we just use one observation series, the "close" price of MBB, and we choose two, three or four states. In the two-state HMM, we assume that the observation data have two states (or regimes): growth and regression. In three-state HMM we assume that an observation sequence has three states: growth, moderate, and regression. In four-state HMM, we suppose an observation sequence has four states: strong growth, weak growth, weak regression, and regression. The HMMs that we use in this section are continuous and for a single observation sequence. Now, let's explore how to use the two-state HMM to find the hidden state sequence for the MBB's close price. Suppose the observation data, O, is the MBB's daily prices from March 16, 2007, to October 26, 2016. We choose initial parameters $\lambda = \{\mathbf{A}, \mu, \sigma, \mathbf{p}\}$ for the Baum-Welch algorithm for N state HMM as:

$$a_{ij} = \frac{1}{N}, \; 1 \leq i, j \leq N \tag{4.1}$$

$$\mu_i = m, \; 1 \leq i \leq N \tag{4.2}$$

$$\sigma_i = s.\epsilon_i, \; 1 \leq i \leq N \tag{4.3}$$

$$\mathbf{p} = (1, 0, ..., 0), \tag{4.4}$$

where m and s are the mean and standard deviation of the observation sequence O, respectively, and ϵ is a random number from the standard normal distribution.

Researchers can choose initial parameters for the Baum-Welch in different ways as long as they satisfy the conditions of matrix $\mathbf{A}$ presented in (1.17) and vector $\mathbf{p}$ presented in (1.19). First, we use the Baum-Welch algorithm with two states, $N = 2$, to calibrate HMM parameters, $\lambda^* = \{\mathbf{A}, \mu, \sigma, \mathbf{p}\}$. Below is the R code for the task.

```
# Calibrate HMM parameters using MBB financial data
# Load required functions
source("C:/Users/scaled.forward.back.R")
source("C:/Users/Cont.Estep.R")
source("C:/Users/Cont.Mstep.R")
source("C:/Users/Cont.BaumWelch.R")
source("C:/Users/Initial.R")
# Load input data
```

```
MBB <- read.table("C:/Users/MBB-D.txt", header = TRUE, sep = "|",
                  as.is = TRUE)
x <- MBB$Close
# Initialize parameters for a continuous HMM with 2 states
Par <- Initial(x, 2)
# Calibrate HMM parameters using the Baum-Welch algorithm
l <- Cont.BaumWelch(x, Par$A, Par$p, Par$mean, Par$sigma)
# Print results
print(l)
```

The Baum-Welch algorithm gives us the following results:

- Transition matrix

$$\mathbf{A} = \begin{bmatrix} 0.9932 & 0.0068 \\ 0.0147 & 0.9853 \end{bmatrix}$$

- Means and variances of two observation probability distributions for two states:

$$\mu_1 = 108.54, \ \sigma_1 = 0.97 \tag{4.5}$$

$$\mu_2 = 103.54, \ \sigma_2 = 2.40 \tag{4.6}$$

- Initial probability of being in state 0 or 1 at time 0:

$$\mathbf{p} = (0, 1).$$

After calibrating the HMM parameters, we have to define the two states based on the results. In general, stocks during the growth state will have higher returns and lower volatility than those during the regression state. However, the two conditions may not happen at the same time, so we define state 0 and state 1 based on the ratio of the means and standard deviations of the two normal distributions.

We define state 1 as the state that has a lower ratio $\frac{\mu_i}{\sigma_i}$, where $i = 1, 2$, and state 0 as the state that has a higher ratio. In this case, we have $\mu_1 > \mu_2$ and $\sigma_1 < \sigma_2$. Thus, we choose state 0 as the state that corresponds with normal distribution with mean μ_1 and standard deviation σ_1, and state 1 as the state that corresponds with normal distribution with mean μ_2 and standard deviation σ_2.

We use the Viterbi algorithm with the calibrated parameters to find "the fittest" hidden states of the observation data. The results are shown in Figure 4.1. The completed codes to find the hidden states of the MBB data are listed below.

```
# Hidden state estimation using MBB financial data and continuous
# HMM
# Load required functions
source("C:/Users/scaled.forward.back.R")
source("C:/Users/Cont.Estep.R")
source("C:/Users/Cont.Mstep.R")
source("C:/Users/Cont.BaumWelch.R")
source("C:/Users/Initial.R")
source("C:/Users/argmin.R")
source("C:/Users/Log.Viterbi.R")
# Load input data
```

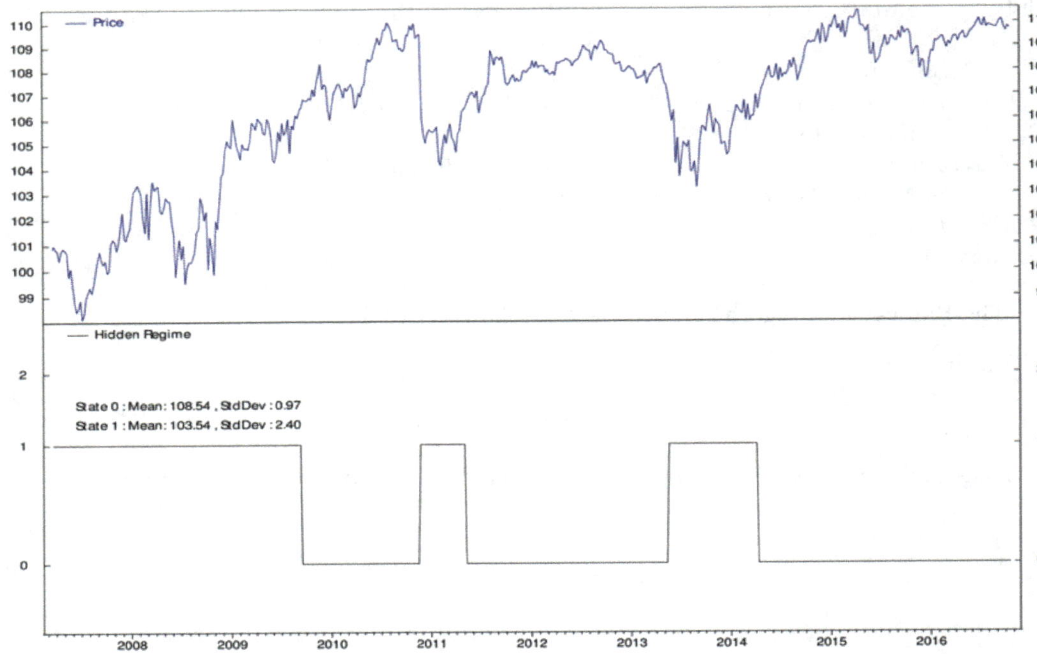

FIGURE 4.1
Finding two states of MBB daily prices using the Viterbi algorithm.

```
MBB <- read.table("C:/Users/MBB-D.txt", header = TRUE, sep = "|",
                  as.is = TRUE)
x <- MBB$Close
# Initialize parameters for a continuous HMM with 2 states
Par <- Initial(x, 2)
# Calibrate HMM parameters using Baum-Welch algorithm
l <- Cont.BaumWelch(x, Par$A, Par$p, Par$mean, Par$sigma)
# Decode hidden states using log-space Viterbi algorithm
Q <- Log.Viterbi(l$A, l$mean, l$sigma, l$p, x)
# Plot hidden states vs. observed data
time <- as.Date(MBB$Date, "%m/%d/%Y")
par(mfrow = c(2, 1))
plot(time, x, type = "l", col = "blue", xlab = "",
     ylab = "MBB Prices")
plot(time, Q, type = "l", col = "red", xlab = "",
     ylab = "Hidden States")
```

Finally, we can see from Figure 4.1 that the HMM captures the hidden states of observation data. MBB price was in state 1 (low returns and high volatility) during the economic recession time 2007-2009, the first quarter of 2011 and from the second quarter of 2013 to the first quarter of 2014.

Similarly, the results of parameters $c\lambda^* = \{\mathbf{A}, \mu, \sigma, \mathbf{p}\}$ calibrated for the two state HMM using weekly data are:

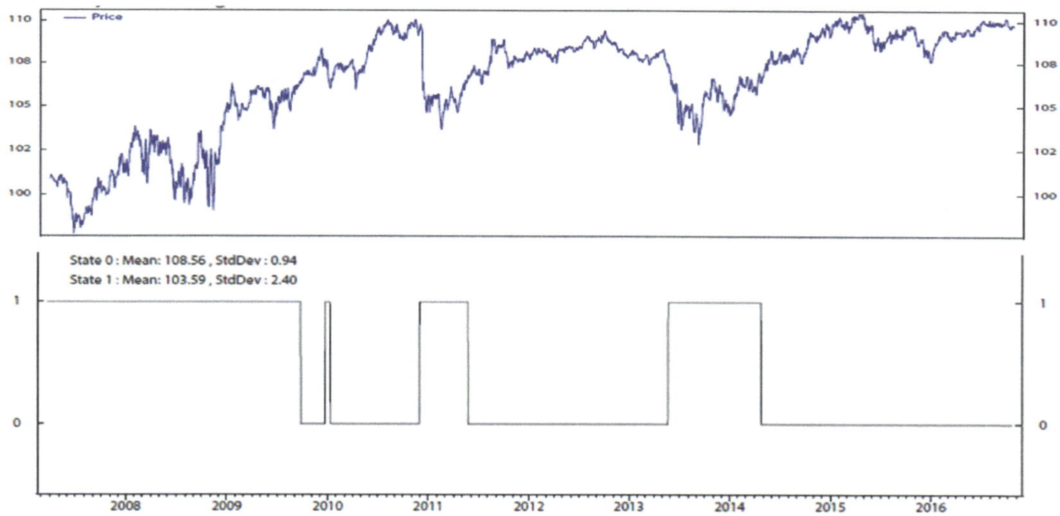

FIGURE 4.2
Finding two states of MBB weekly prices using the Viterbi algorithm.

- Transition matrix

$$\mathbf{A} = \begin{bmatrix} 0.9932 & 0.0068 \\ 0.0147 & 0.9853 \end{bmatrix}$$

- Means and variances of two observation probability distributions for two states:

$$\mu_1 = 108.56, \ \sigma_1 = 0.94 \quad (4.7)$$

$$\mu_2 = 103.59, \ \sigma_2 = 2.40 \quad (4.8)$$

- Initial probability of being in state 0 or 1 at time 0:

$$\mathbf{p} = (0, 1).$$

The two states simulated of the HMM using the monthly MBB are presented in Figure 4.2. Results of parameters $\lambda^* = \{\mathbf{A}, \mu, \sigma, \mathbf{p}\}$ calibrated for the two state HMM using monthly data are:

- Transition matrix

$$\mathbf{A} = \begin{bmatrix} 0.9687 & 0.0313 \\ 0.0619 & 0.9381 \end{bmatrix}$$

- Means and variances of two observation probability distributions for two states:

$$\mu_1 = 108.72, \ \sigma_1 = 0.98 \quad (4.9)$$

$$\mu_2 = 103.94, \ \sigma_2 = 2.41 \quad (4.10)$$

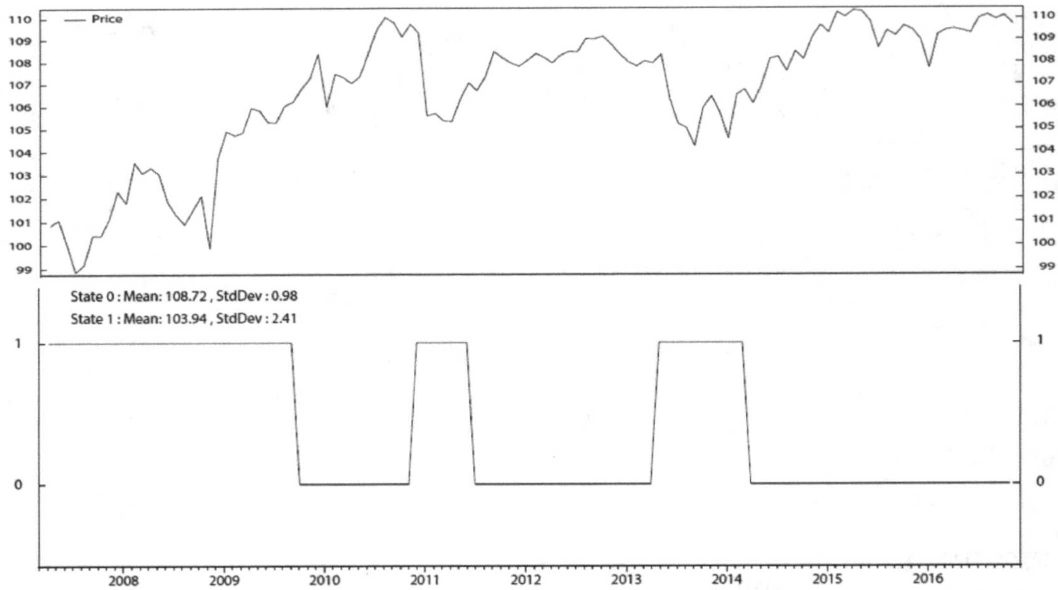

FIGURE 4.3
Finding two states of MBB monthly prices using the Viterbi algorithm.

- Initial probability of being in state 0 or 1 at time 0:

$$\mathbf{p} = (0, 1).$$

The two states simulated of the HMM using the monthly MBB are presented in Figure 4.3. From Figures 4.1–4.3, we can see that the two state HMM detects the "Bear" periods, or the recession periods, constantly for the daily, weekly, or monthly MBB's prices. The "Bear" periods are from mid 2007-to mid 2010, the end of 2011 to mid 2012, and 2013-2014.

Similar to the two-state HMM, we have the following results of the three-state and four-state HMM.

Results of using three-state HMM to calibrate the model's parameters and find hidden states of MBB observation:

- Transition matrix A:

$$\mathbf{A} = \begin{bmatrix} 0.9565 & 0.0356 & 0.0079 \\ 0.0070 & 0.9978 & 0.0152 \\ 0.0063 & 0.0258 & 0.9979 \end{bmatrix}$$

- Means and variances of three observation probability distributions (three normal distributions):

$$\mu_0 = 108.81, \sigma_0 = 0.78 \tag{4.11}$$
$$\mu_1 = 105.90, \sigma_1 = 1.01 \tag{4.12}$$
$$\mu_2 = 101.13, \sigma_2 = 1.73 \tag{4.13}$$

- Initial probability of being in state 0, 1, or 2 at time 0:

$$\mathbf{p} = (0, 0, 1).$$

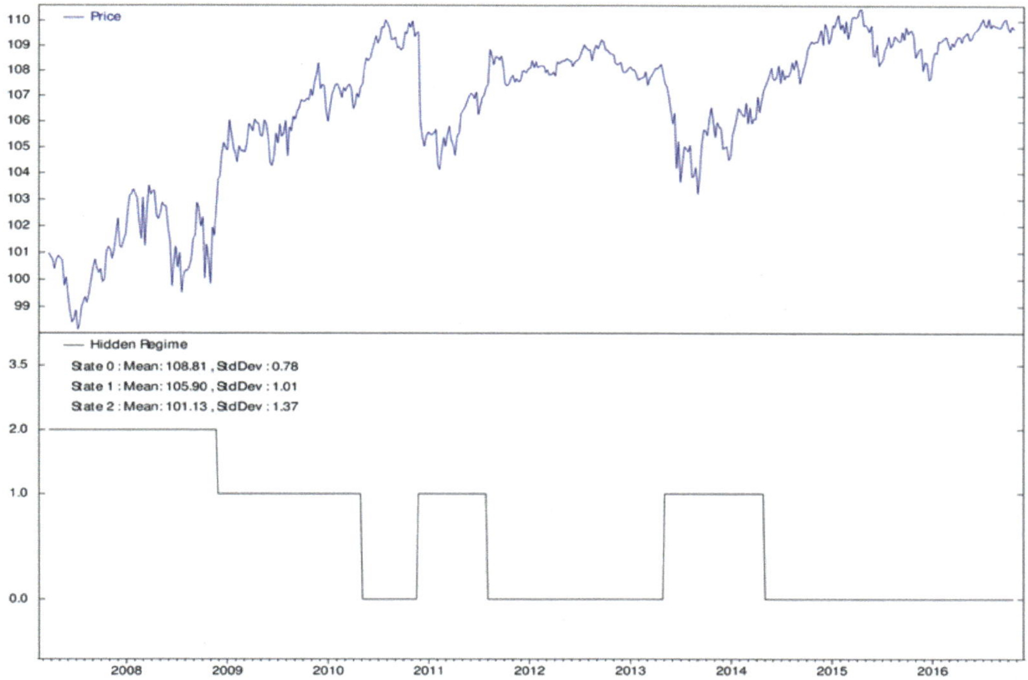

FIGURE 4.4
Finding three states of MBB daily prices using the Viterbi algorithm.

The hidden sequence of the observations using the three-state HMM is presented in Figure 4.4. Results of using four-state HMM to calibrate the model's parameters and find hidden states of MBB observation:

- Transition matrix A:

$$\mathbf{A} = \begin{bmatrix} 0.6885 & 0.2919 & 0 & 0.0196 \\ 0.7251 & 0.2677 & 0.0072 & 0 \\ 0.0087 & 0 & 0.8594 & 0.1119 \\ 0 & 0 & 1 & 0 \end{bmatrix}$$

- Means and variances of four observation probability distributions (four normal distributions):

$$\mu_0 = 109.39, \sigma_0 = 0.47 \tag{4.14}$$
$$\mu_1 = 107.79, \sigma_1 = 0.55 \tag{4.15}$$
$$\mu_2 = 105.46, \sigma_2 = 0.78 \tag{4.16}$$
$$\mu_3 = 101.12, \sigma_3 = 1.37 \tag{4.17}$$

- Initial probability of being in state 0 to 3 at time 0:

$$\mathbf{p} = (0, 0, 0, 1)$$

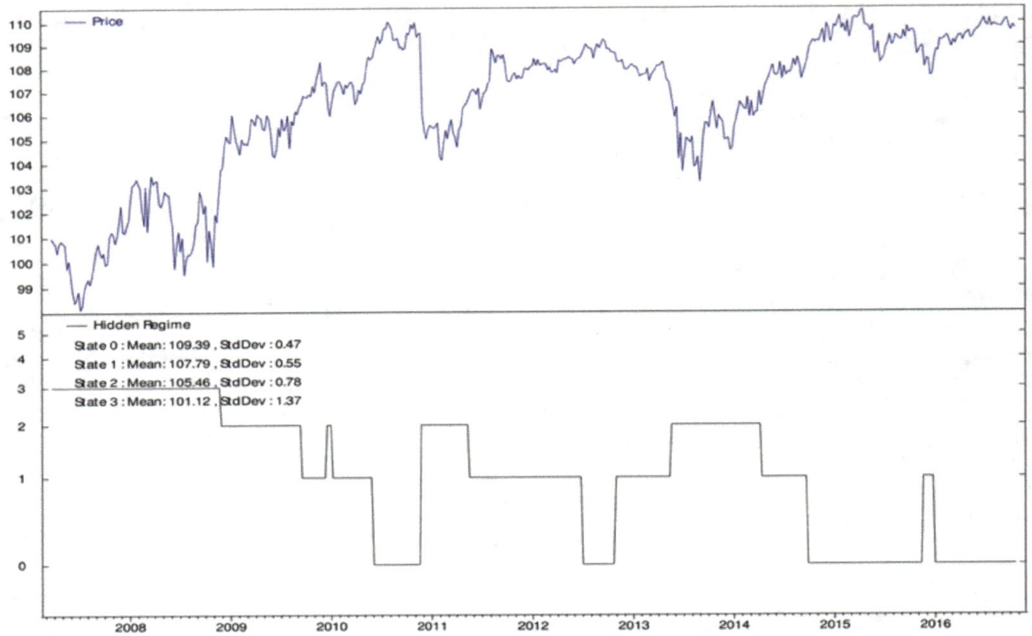

FIGURE 4.5
Finding four states of MBB daily prices using the Viterbi algorithm.

The "best fit" hidden state sequence using the four-state HMM is displayed in Figure 4.5. The results show that we can use the HMM to find hidden states of historical observation data. Furthermore, we can use the HMM to predict the probability of being in different states in the future. We present this technique in the next section.

4.3.4 Using an HMM to Estimate Recession Probabilities

In this section, we use an HMM to predict the probability of MBB being in the recession state. We use the HMM for both single observation data and multiple observation data in this section. For single observation data we use MBB "close" price, and for multiple observation data we use MBB "close", "open", "low", and "high" prices.

To make predictions and model validations, we use two-year MBB daily data from October 16, 2014, through October 26, 2016. We used fixed-length windows of one-year data, with each window having 252 consecutive trading days, to predict the probability of MBB being in recession the next day. The first data window, from October 30, 2014, to October 27, 2015, was used to predict the probability of being in the recession state on October 28, 2015.

We first use the Baum-Welch algorithm to calibrate HMM parameters and then define states 0 and 1 based on the calibrated parameters of the two normal distributions: μ, σ. The recession state would have a lower ratio $\frac{\mu}{\sigma}$ compared to those of the "Bull" market, or the growing market. Using the defined states and the first-order property of the Markov chain of HMM state sequences, we can predict the probability of being in state 1 by using the $\xi_t(i,j)$ function. The observation probability was calculated by using transition matrix A and the normal probability distribution corresponding to the recession state. After the

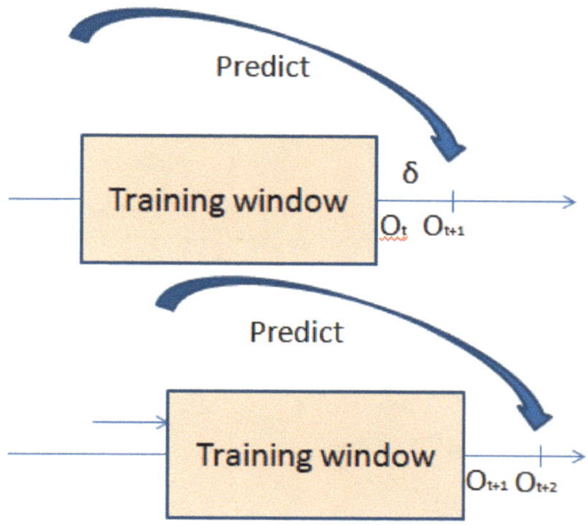

FIGURE 4.6
Moving window technique for probability prediction.

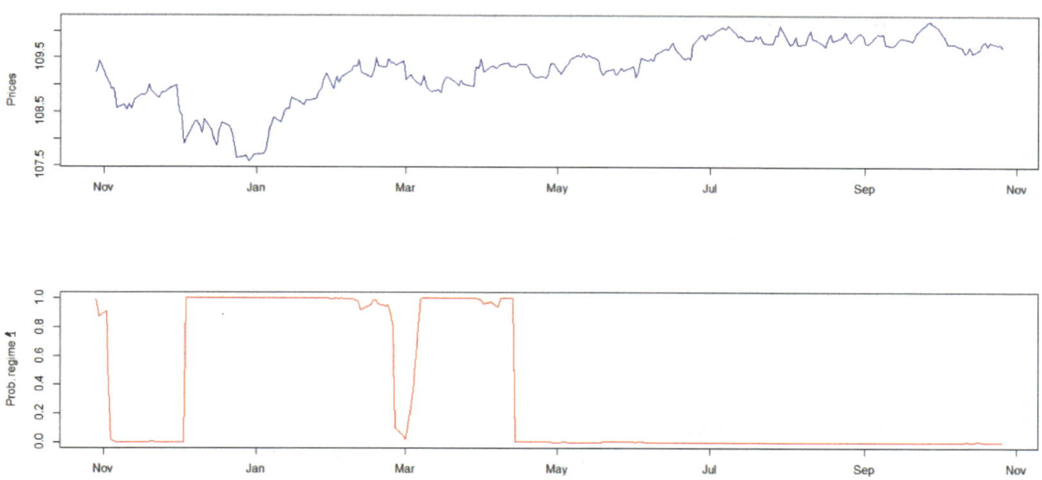

FIGURE 4.7
Predicted probability of being in the "Bear" market (state 1) using one observation sequence.

first prediction, we moved the training window up one day and used the new data set to make the second prediction. The moving window technique is presented in Figure 4.6. The predicted probability of being in the recession state for a one-year period is presented in Figure 4.7 and Figure 4.8. In Figure 4.7 we use one observation sequence, the "close" price of MBB, and in Figure 4.8 we use four observation sequences: "close," "open," "low" and "high" price. We can see that the predicted probability of being in the "Bear" state using four observation sequences is more sensitive with the MBB prices compared to using the one observation sequence. Therefore, in the next section, we will use four observation sequences to predict the mortgage backed security ETFs prices. The completed codes for predicting the probability of being in the recession state of the MBB daily data were presented below.

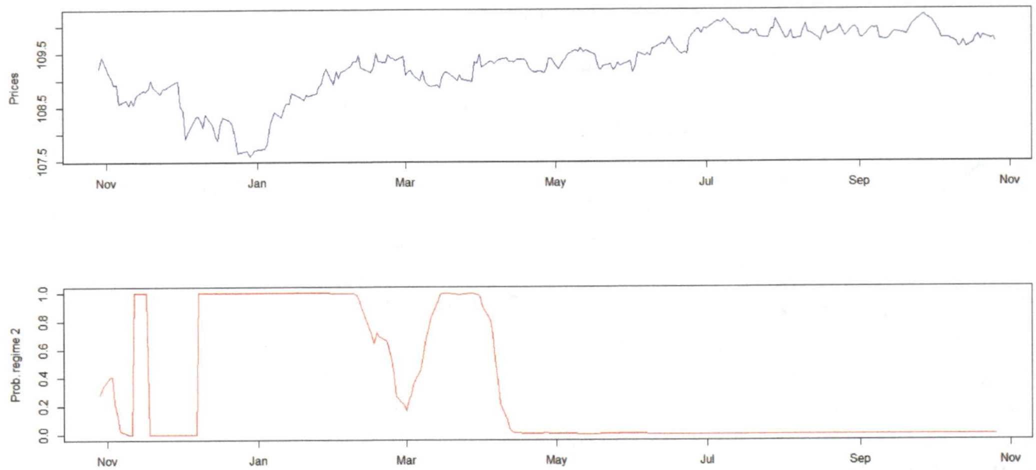

FIGURE 4.8
Predicted probability of being in the "Bear" market (state 1) using four observation sequences.

```
# Predict recession probabilities using a rolling-window HMM with
# multiple observation sequences
# Load required functions
source("C:/Users/BW.NdataR")
source("C:/Users/ESTEP.N.DATA.R")
source("C:/Users/scaled.forward.back.R")
source("C:/Users/Initial.R")
source("C:/Users/Estep.Ndata.R")
# Load observation data
DJ <- read.table("C:/Users/MBB-D.txt", header = TRUE, sep = "|",
                 as.is = TRUE)
# Set parameters
WD <- 252                        # Window length
D <- length(DJ$Date) - WD        # Starting index for rolling window
K <- 4                           # Number of observation sequences
N <- 2                           # Number of hidden states
Pstate2 <- rep(0, WD)            # Store predicted probabilities
# Rolling-window prediction loop
for (i in 1:WD) {
  y1 <- as.numeric(DJ$Close[(D - WD + i + 1):(D + i)])
  y2 <- as.numeric(DJ$Open[(D - WD + i):(D + i - 1)])
  y3 <- as.numeric(DJ$Low[(D - WD + i + 1):(D + i)])
  y4 <- as.numeric(DJ$High[(D - WD + i):(D + i - 1)])
  T <- WD
  # Construct observation array
  y <- array(NA, dim = c(K, T, 1))
```

```
  y[1, , ] <- y1
  y[2, , ] <- y2
  y[3, , ] <- y3
  y[4, , ] <- y4
  # Initialize or update HMM parameters
  if (i == 1) {
    Par <- hmm.iniNdata(y, N)
    k <- BW.Ndata(y, Par$A, Par$p, Par$mean, Par$sigma, T, K, N)
  } else {
    k <- BW.Ndata(y, k$A, k$p, k$mean, k$sigma, T, K, N)
  }
  # Determine recession state based on signal-to-noise ratio
  TT <- (k$mean[1] / k$sigma[1] < k$mean[2] / k$sigma[2])
  Pstate2[i] <- if (TT) k$u[T, 1] else k$u[T, 2]
}
# Print prediction window
print("Begin date")
print(DJ$Date[D + 2])
print("End date")
print(DJ$Date[D + WD])
# Plot results
y11 <- DJ$Close[(D + 2):(D + WD)]
State2 <- round(Pstate2)
time <- as.Date(DJ$Date[(D + 2):(D + WD)], "%m/%d/%Y")
par(mfrow = c(2, 1))
plot(time, y11, type = "l", col = "blue", xlab = "",
     ylab = "Prices")
plot(time, State2[1:(WD - 1)], type = "l", col = "red", xlab = "",
     ylab = "Prob. regime 2")
```

4.3.5 Using an HMM for Price Prediction and MBS ETF Trading

The accuracy of a prediction model is the most important factor. The HMM will perform differently based on the number of hidden states. Therefore, to forecast the MBB's prices using the HMM, we have to choose the number of states first. In this section, we use the two standard criteria—the Akaike information criterion, AIC, (Akaike, 1974) and the Bayesian information criterion, BIC, (Schwarz, 1978)—to choose the best model among the HMM with two, three, or four states. Then, we will use the HMM with two, three or four states to predict MBB's prices.

4.3.5.1 Model Selection

Choosing a number of hidden states for an HMM is a critical task. We use two common criteria, the AIC and the BIC, to evaluate the performances of the HMM with different numbers of states. This criteria measures how well a model fits with given data by evaluating the log likelihood of the model. The likelihood, L, of an HMM is the probability of observations given the model's parameters:

$$L = \mathbb{P}(O|\lambda).$$

The observation data O is multiple observation sequences with dimensions 4 by n, 4 observation sequences and each sequence has n data points. In this section we used the whole

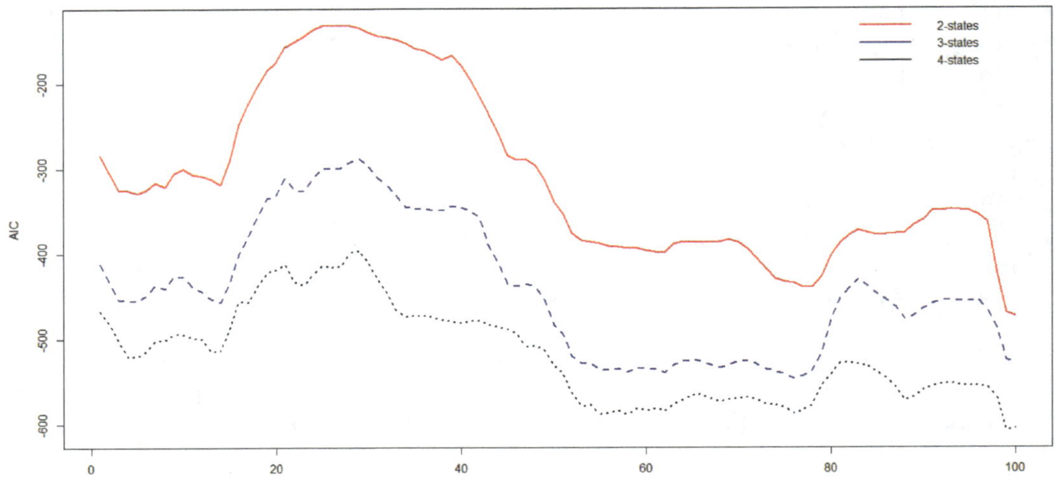

FIGURE 4.9
AIC values of training the HMM model with different number of states using four observation sequences.

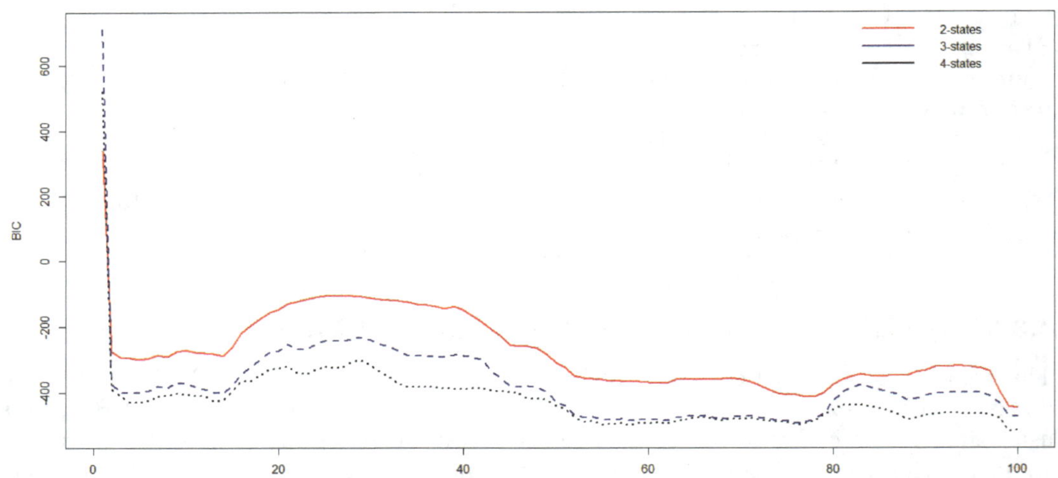

FIGURE 4.10
BIC values of training the HMM model with different number of states using four observation sequences.

data set which is daily MBB prices from 1/1/2009 to 10/26/2016. Thus $n = 1950$ observation points for each observation data sequence: "open", "close", "high", "low". We used the moving training window that was described in Section 4.3.4 to make calibrations and plot the results in Figure 4.9 and Figure 4.10. The length of each training window is 100 observation points for each sequence. Recall that the model with a lower AIC or BIC is better. Both the AIC and BIC results in Figure 4.9 and Figure 4.10 have the lowest values for the HMM with four states. Therefore, we can conclude that using the HMM with four states to predict the MBB prices will have more accurate predictions than using the model with two or three states. To test this assumption, in the next section we will use the HMM with two, three, and four states to predict MBB prices and compare the predicted errors to see if the results are consistent with the AIC or BIC test results. The following is the code to find the optimal number of states for HMMs by calculating their AIC and BIC, using MBB's daily prices.

```
# Evaluate optimal number of HMM states using AIC and BIC on MBB
# financial data
# Load required functions
source("C:/Users/BW.NdataR")
source("C:/Users/ESTEP.N.DATA.R")
source("C:/Users/scaled.forward.back.R")
source("C:/Users/Initial.R")
source("C:/Users/Estep.Ndata.R")
# Load observation data
Sp <- read.table("C:/Users/MBB-D.txt", header = TRUE, sep = "|",
                 as.is = TRUE)
T1 <- length(Sp$Date)
K <- 4  # Number of observation sequences
# Construct observation array
y <- array(NA, dim = c(K, T1, 1))
y[1, , ] <- as.numeric(Sp$Close)
y[2, , ] <- as.numeric(Sp$Open)
y[3, , ] <- as.numeric(Sp$Low)
y[4, , ] <- as.numeric(Sp$High)
# Set up rolling window
WD <- 100
T2 <- T1 - WD
T <- WD
# Initialize AIC and BIC storage
AIC <- array(NA, dim = c(3, WD))
BIC <- array(NA, dim = c(3, WD))
# Train HMMs and compute AIC/BIC for N = 2, 3, 4
for (N in 2:4) {
  for (l in 1:WD) {
    x <- array(NA, dim = c(K, WD, 1))
    for (j in 1:K) {
      x[j, , ] <- y[j, , ][(T2 - WD + 1):(T2 + l - 1)]
    }
    # Initialize or update parameters
    if (l == 1) {
      Par <- Initial(x, N)
      k <- BW.Ndata(x, Par$A, Par$p, Par$mean, Par$sigma, T, K, N)
    } else {
      k <- BW.Ndata(x, k$A, k$p, k$mean, k$sigma, T, K, N)
    }
    # Compute AIC and BIC
    b <- N * N + 2 * N - 1
    AIC[N - 1, l] <- -2 * k$LL + 2 * b
    BIC[N - 1, l] <- -2 * k$LL + b * log(4 * WD)
  }
```

```
}
# Plot AIC
y1 <- cbind(AIC[1, ], AIC[2, ], AIC[3, ])
matplot(y1, type = "l", col = c("red", "blue", "black"), lwd = 2,
        xlab = "Simulations", ylab = "AIC",
        main = "AIC of HMM for MBB Daily Data")
legend("topright", legend = c("2-states", "3-states", "4-states"),
       col = c("red", "blue", "black"), lwd = 2, bty = "n")
# Plot BIC
y2 <- cbind(BIC[1, ], BIC[2, ], BIC[3, ])
matplot(y2, type = "l", col = c("red", "blue", "black"), lwd = 2,
        xlab = "Simulations", ylab = "BIC",
        main = "BIC of HMM for MBB Daily Data")
legend("topright", legend = c("2-states", "3-states", "4-states"),
       col = c("red", "blue", "black"), lwd = 2, bty = "n")
```

4.3.5.2 Price Prediction

In this section, we use the HMM and historical price data for MBB to predict its future prices and compare the predictions with the market prices. We forecast MBB prices using the HMM with different numbers of states and calculate the mean absolute percentage errors of the estimates,

$$MAPE = \frac{1}{n}\sum_{i=1}^{n}\frac{|M_i - P_i|}{M_i}, \tag{4.18}$$

where n is the number of predicted points, M_i is market price and P_i is the predicted price of a stock, $i = 1, ..., n$. We use the HMM with multiple observation sequences ("open", "low", "high" and "close" price) to predict future "close" price.

To find this prediction, we divided MBB daily data from March 16, 2007 to October 26, 2016 into three parts: one part for back tracking , one part for model training and one part for model validating. We use data from a one-year time period (252 trading days) for a training window, and for convenience, we use one-year data, from October 28, 2015, to October 26, 2016, for model testing.

The prediction process can be divided into three steps:

1. Calibrate the HMM parameters and calculate the likelihood of the model,
2. Find a day in the past that has a similar likelihood to that of the recent day,
3. Use the difference of stock prices on the "similar" day in the past to predict future stock prices.

This prediction approach is based on the work of Hassan and Nath [61]. In order to predict the ETF price for October 28, 2015, we use a one-year data window from October 30, 2014 to October 27, 2015.

In the first step, we use the Baum-Welch algorithm to calibrate the parameters for the HMM and use the forward algorithm to calculate the probability of observation for the data set.

In the second step, we move the training data backward by one day, which is called a tracking data window, and use calibrated parameters, λ, to calculate the probability of the tracking data window. We continue to move the tracking window backward until we find the "similar" tracking data window that has a very close probability of observation with the training data window (the chosen error is 10^{-6}).

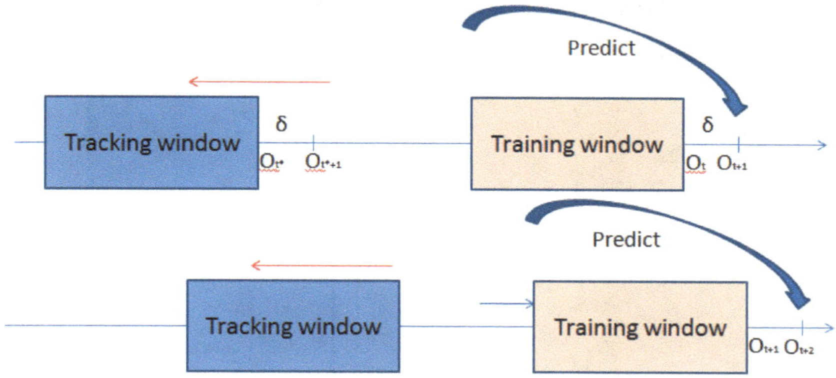

FIGURE 4.11
MBB price prediction process.

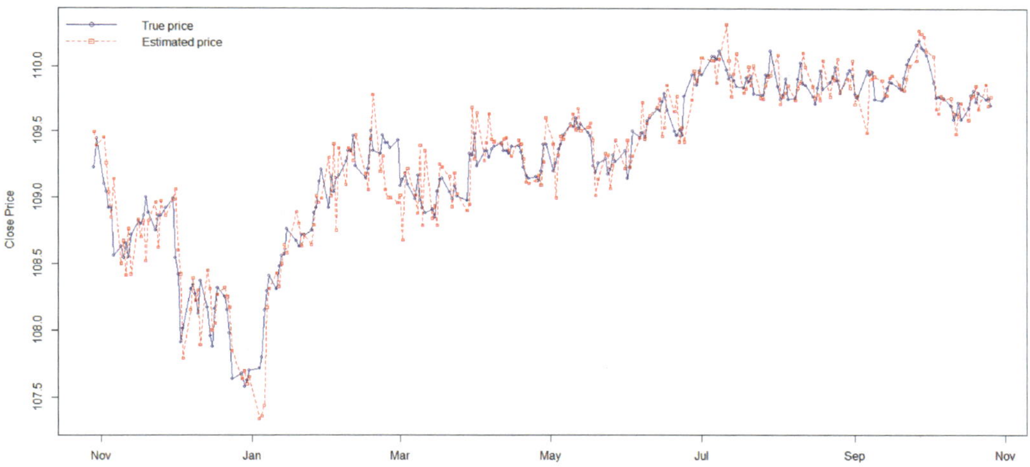

FIGURE 4.12
MBB price prediction from October 28, 2015 to October 26, 2016, using two state HMM.

In the last step, we use the difference, Δ, between the prices on the end date of the "similar" tracking window and the next consecutive day to predict the price of MBB for October 28, 2015. We assume that the change of MBB's price in October 27, 2016, and October 28, 2016, is equal to Δ. Thus, the predicted price equals the price on October 27, 2016 plus Δ. For the second prediction, we move the training window up one day and repeat the process. The prediction process is presented in Figure 4.11. The results of price predictions with two-, three- and four-state HMMs are presented in Figures 4.12 - 4.14. We can see from the figures that the HMM with three or four-states is more accurate than the HMM with two-states in stock price predictions.

We calculate the mean absolute percentage errors, MAPE, of the predictions using these three models and present the results in Table 4.2.

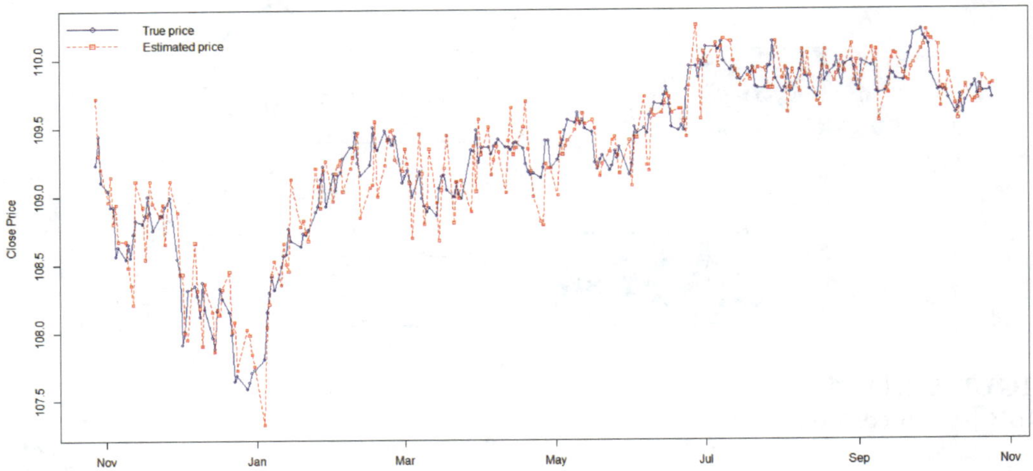

FIGURE 4.13
MBB price prediction from October 28, 2015 to October 26, 2016, using a three state HMM.

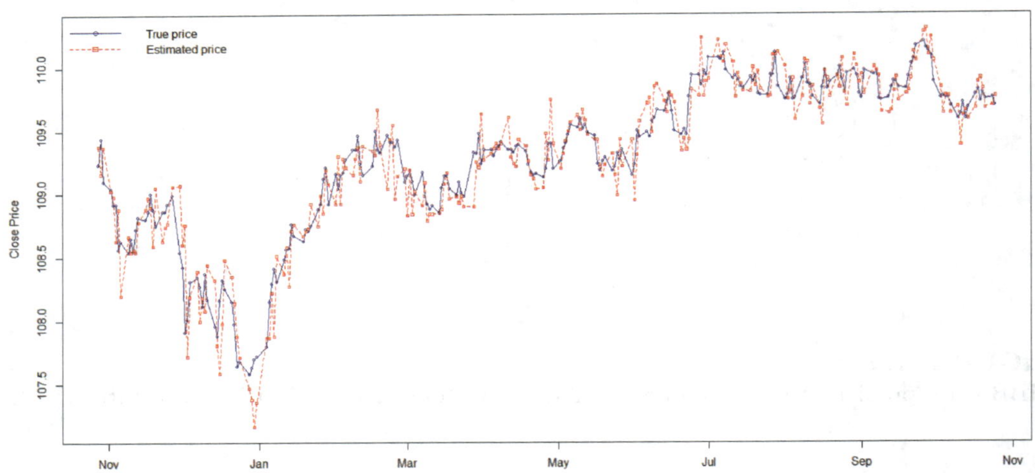

FIGURE 4.14
MBB price prediction from October 28, 2015 to October 26, 2016, using a four state HMM.

The results in Table 4.2 show that the four-state HMM gave the smallest error for predicting MBB's prices. These result are consistent with the results from Section 4.3.5.1. Specifically, the AIC and BIC results indicated that the HMM is most accurate with four states.

TABLE 4.2
MBB price prediction errors using the HMM with different numbers of states

Number of States	2	3	4
MAPE	0.001291	0.001316	0.001239

The following code predicts the MBB weekly price using an HMM with four states, based on input data consisting of four observation sequences: “close”, “open”, “high”, and “low” prices of MBB on a weekly basis.

```
# Predict MBB prices using HMM with multiple observation sequences
# Load required functions
source("C:/Users/BW.NdataR")
source("C:/Users/ESTEP.N.DATA.R")
source("C:/Users/scaled.forward.back.R")
source("C:/Users/Initial.R")
source("C:/Users/Estep.Ndata.R")
source("C:/Users/argmin.R")
# Load observation data
Sp <- read.table("C:/Users/MBB-W.txt", header = TRUE, sep = "|",
                 as.is = TRUE)
T1 <- length(Sp$Date)
K <- 4   # Number of observation sequences
N <- 4   # Number of hidden states
# Construct observation array
y <- array(NA, dim = c(K, T1, 1))
y[1, , ] <- as.numeric(Sp$Close)
y[2, , ] <- as.numeric(Sp$Open)
y[3, , ] <- as.numeric(Sp$Low)
y[4, , ] <- as.numeric(Sp$High)
# Set up prediction window
WD <- 52
T2 <- T1 - WD
T <- WD
price <- numeric(WD)
# Rolling prediction loop
for (l in 1:WD) {
  x <- array(NA, dim = c(K, WD, 1))
  for (j in 1:K) {
    x[j, , ] <- y[j, , ][(T2 - WD + 1 + 1):(T2 + 1)]
  }
  # Initialize or update HMM parameters
  if (l == 1) {
   Par <- hmm.iniNdata(x, N)
   k <- BW.NdataNew(x, Par$A, Par$p, Par$mean, Par$sigma, T, K, N)
  } else {
   k <- BW.NdataNew(x, k$A, k$p, k$mean, k$sigma, T, K, N)
  }
  # Find most similar past day based on log-likelihood
  ll <- numeric(WD)
```

```
  for (i in 1:WD) {
    for (j in 1:K) {
      z <- y[j, , ][(T2 - WD - i + 1 + 1):(T2 - i + 1)]
      ll[i] <- ll[i] + forward(z, k$A, k$p, k$mean, k$sigma)
    }
  }
  point <- argmin(abs(ll - k$LL))
  # Predict price using matched historical difference
  price[1] <- y[1, , ][T2 + 1] + (y[1, , ][T2 - point + 1 + 1]
              - y[1, , ][T2 - point + 1])
}
# Print prediction window
print("Begin date")
print(Sp$Date[T2 + 2])
print("End date")
print(Sp$Date[T1])
# Compute relative error
true.price <- y[1, , ][(T2 + 2):T1]
pricenew <- price[1:(WD - 1)]
RER <- mean(abs(pricenew - true.price) / true.price)
print("Error is")
print(RER)
# Plot results
g_range <- range(min(true.price), true.price, pricenew)
time <- as.Date(Sp$Date[(T1 - WD + 2):T1], "%m/%d/%Y")
plot(time, true.price, type = "o", cex = 0.5, col = "blue",
     xlab = "", ylab = "Close Price", ylim = g_range)
lines(time, pricenew, type = "o", pch = 22, cex = 0.5,
    col = "red", lty = 2)
legend("topleft", legend = c("True price", "Estimated price"),
    col = c("blue", "red"), pch = 21:22, lty = 1:2, bty = "n")
```

Readers can modify the codes to accommodate different input data. The R code can also be applied to the following sections. Therefore, we will not include the codes for the remaining sections of this chapter.

4.3.5.3 Model Validations

We now compare the HMM with the standard forecast model, the historical average model for stock return predictions. Among the eight MBS ETFs listed in Table 1, we can choose only five that have long enough historical data to implement the HMM. The five MBS ETFs are MBB, GNMA, CMBS, VMBS and MBG. We use daily and weekly prices of the ETFs. Due to the limitations of data, we use a training window of 52 weeks for weekly data (one-year period from November 2, 2015 to October 24, 2016) and training windows of 100 days (from June 7, 2016 to October 26, 2016) and 252 days (from October 28, 2015 to October 26, 2016) for daily data. For convenience, the length of the out-of-sample predictions equals the length of the training window. That means we predict prices for 52 weeks, 100 days and 252 days. For each ETF, we use its four observation sequences—the "open", "low", "high" and "close" prices—to predict the close price in the future time. With a training window

TABLE 4.3
Errors of out-of-sample MBS ETFs predicted prices using the HAM vs. the HMM

ETF	Model	100 Days (6/7/2016-10/26/2016)	252 Days (10/28/2015-10/26/2016)	52 Weeks (11/2/2015-10/24/2016)
MBB	HAM	0.0007	0.0009	0.0018
	HMM	0.0005	0.0009	0.0014
BNMA	HAM	0.0013	0.0016	0.0021
	HMM	0.0009	0.0016	0.0020
VMBS	HAM	0.0009	0.0010	0.0017
	HMM	0.0007	0.0011	0.0013
CMBS	HAM	0.0021	0.0022	0.0047
	HMM	0.0017	0.0023	0.0038
MBG	HAM	0.0030	0.0016	0.0027
	HMM	0.0027	0.0014	0.0018

size m, the historical average model (HAM) is defined as

$$r_{m+t} = \frac{1}{m+t-1} \sum_{i=1}^{m+t-1} r_i. \tag{4.19}$$

Since the moving average model only works well for a stationary series, we use r_t is as the total return of stock at time t, $r_t = P_t - P_{t-1}$, where P_t is price at time t of a stock. After using HAM to predict return, we use the formula $P_t = P_{t-1} + r_t$ to obtain the predicted price of the stock. The MAPE of the estimates, as defined in Section 3.4, is used to compare the performances of the HMM and the HAM,

$$MAPE = \frac{1}{n} \sum_{i=1}^{n} \frac{|M_i - P_i|}{M_i}.$$

Errors of stock price predictions using the two models are presented in Table 4.3. Based on the error estimator, MAPE, the HMM beats the HAM in predicting the ETFs in most cases. In only two cases, forecasting VMBS and CMBS using a 252-day training window, the HAM outperformed the HMM. However, by definition, the MAPE measures the mean of absolute percentage errors. It does not tell us the accuracy of the predictions according to the real observation trends.

One of the disadvantages of the moving average method is that it is not sensitive to a sudden change of the momentum in the prices. Therefore, it is not the desired candidate to predict trading signals or trends of observations. By comparison, the HMM is based on regimes of observation sequences and the probability of jumping from one state to another. Thus, the HMM is a good model to capture the regime shifts of observations. In the next section we investigate the efficiencies of the HMM and the HAM in predicting trading or trending signals.

4.3.5.4 Trading Mortgage-Backed Securities

We now apply the two models, the HMM and the HAM, to trading the MBS ETFs based on the predicted returns. Similar to Section 4.3.5.2, we will trade these five ETFs—MBB, GNMA, CMBS, VMBS and MBG—in three different periods: 100 days, 252 days for daily trading, and 52 weeks for weekly trading. We assume no transaction fee when we buy or sell the ETFs by the close prices, and we buy or sell 100 shares in each trade. This trading strategy is based on the predicted prices from Section 4.3.5.2. If the model predicts that the price of the ETF goes up, we will buy it and hold until the model predicts that the price goes down. Results of the trading are summarized in Table 4.4. The table gives the earning percentage in the whole period of trading using the HMM and the HAM.

TABLE 4.4
Comparison of trading MBS ETFs using HAM vs. HMM.

ETF	Model	100 Days	252 Days	52 Weeks
		(6/7/2016-10/26/2016)	(10/28/2015-10/26/2016)	(11/2/2015-10/24/2016)
MBB	HAM	0.23%	-0.91%	0.03%
	HMM	1.96%	2.42%	2.12%
BNMA	HAM	0.24%	-0.47%	-0.75%
	HMM	3.98%	4.73%	2.81%
VMBS	HAM	0.28%	-0.86%	-0.02%
	HMM	2.33%	4.39%	2.17%
CMBS	HAM	0.17%	-0.95%	0.29%
	HMM	3.53%	9.62%	5.35%
MBG	HAM	-0.07%	0.00%	0.00%
	HMM	2.18%	7.61%	3.91%

Based on the results in Table 4.4, we see that the HMM captures price changes well. Therefore, the HMM yields a higher return than the moving average model in trading the ETFs and the results show that the HMM is a promising model for stock trading.

4.3.6 Conclusions

In Section 4.3, we presented an introduction of the HMM: the historical developments of the model, its basic elements and algorithms, and applications of the model in finance and actuarial fields. We presented simple examples to explain the model's concepts and three main algorithms. We then applied the HMM to mortgage-backed securities exchange-traded funds, MBS ETFs. The Baum-Welch algorithm was used to calibrate the model's parameters, and the Viterbi algorithm was used to find the hidden state sequence of the observation data. We used the HMM with two, three, and four states to predict the probability of MBB, an IShares MBS ETF, being in the recession state, using one observation sequence and four observation sequences. The results showed that the HMM is a good model for forecasting observations' regimes. The more observation sequences we used, the better the predictions were.

Before applying the model for predicting and trading the MBS ETFs, we used the two standard goodness-of-fit tests, the AIC and the BIC, to choose the best HMM (or the number of states for the HMM) for the ETFs. The results showed that the HMM with four states is the best model among the two-, three-, or four-state HMM. Based on the model selection results, we used the four-state HMM with multiple observation sequences to predict the ETFs' prices. We compared the HMM with the HAM in predicting the prices. The results showed that the HMM outperformed the HAM for various training windows and out-of-sample time periods. The trading results based on the predicted prices showed that the HAM was worse for trading because it failed to capture the trading signals or the future trends of the underlying access. By comparison, the HMM worked well in trading the MBS ETFs because it predicted precisely the future returns of the funds. In conclusion, the HMM is a stochastic model that is based on the constant transition matrix of observations' hidden states. It has many applications in finance and actuarial sciences. It is a potential model for predicting extreme regimes of observation sequences and for forecasting prices and trading indexes.

4.4 HMM for Stock Selection

An HMM is typically used to predict the hidden regimes of observation data. Therefore, this model finds applications in many different areas, such as speech recognition systems, computational molecular biology, and financial market predictions. In this section, we use the HMM for stock selection. We first use the HMM to make monthly regime predictions for the four macroeconomic variables: inflation (consumer price index (CPI)), industrial production index (INDPRO), stock market index (S&P 500) and market volatility (VIX). At the end of each month, we calibrate the HMM parameters for each of these economic variables and predict its regimes for the next month. We then look back into historical data to find the time periods for which the four variables had similar regimes with the forecast regimes. Within those similar periods, we analyze all of the S&P 500 stocks to identify which stock characteristics have been well rewarded during the time periods and assign scores and corresponding weights for each of the stock characteristics. A composite score of each stock is calculated based on the scores and weights of its features. Based on this algorithm, we choose the top 50 ranking stocks to buy. We compare the performances of the portfolio with the benchmark index, S&P 500. With an initial investment of $100 in December 1999, over 15 years, in December 2014, our portfolio had an average gain per annum of 14.9% vs. 2.3% for the S&P 500.

4.4.1 Introduction

In many financial problems, the states of a system can be modeled as a Markov chain in which each state depends on the previous state in a non-deterministic way. In the HMM, these states are invisible, while observations (the inputs of the model), which depend on the states, are visible. An observation at time t of an the HMM has a certain probability distribution corresponding to a possible state.

Researchers have applied the HMM for analyzing economic trends. The HMM was used in [26] with three states to predict currency crises in six developing countries. Hassan and Nath [61] used the HMM to forecast the stock price for interrelated markets. Kritzman, Page and Turkington [79] applied the HMM with two states to predict regimes in market turbulence, inflation and the industrial production index. Guidolin, and Timmermann [58] used the HMM with four states and multiple observations to study asset allocation decisions based on regime switching in asset returns. Ang and Bekaert [5] applied a RSM (other name for the HMM) for international asset allocation. Nguyen [103] used the HMM with both single and multiple observations to forecast economic regimes and stock prices. Nobakht, Joseph and Loni [108] implemented the HMM using multiple observation data (open, close, low, and high) prices of a stock to predict its close prices. However, the momentum of a stock depends on many different factors, such as the corporate financial condition and management and the overall economy and industry conditions. These factors and corresponding stock returns vary widely over different macro regimes. In addition, long-term stock investments' returns depend on the trends of these economic factors. Therefore, in this book, we develop a new approach of the HMM: making monthly stock selections based on their historical performances on economic regimes.

We analyze the performances of stocks' returns on each macro regime to make stock selections instead of applying the HMM directly to predict their prices. Our approach differs from [5,79] in that the authors used the HMM to make asset allocations, while we build up a stock portfolio based on macro regimes. We chose four main macroeconomic variables that indeed have significant effects on stock prices: inflation, industrial production index, stock market index, and market volatility. First, we use the HMM to predict regimes for each economic indicator for the next month. Based on the performances of stock characteristics in the past on the similar regimes, we then assign a score and weight for each characteristic. Second, we calculate a composite score of each stock based on its features' scores and weights. Finally, we make a selection of the 50 stocks that had the highest scores among all S&P 500 stocks to add to our portfolio and continue the stock selection process at the end of each month.

4.4.2 Data Selections

After analyzing the performances of the benchmark market index (S&P 500) on two defined regimes of different economic indicators, we found that the S&P 500 performs significantly different across different states of four macroeconomic variables: inflation (consumer price index (CPI)), industrial production index (INDPRO), stock market index (S&P 500), and market volatility (VIX), a measure of market expectations of near-term volatility conveyed by S&P 500 stock index option prices. Thus, we chose these four variables as the four macroeconomic indicators for our model. These are the definitions of the variables:

1. Inflation: we use the 12-month changes (%) in CPI where CPI is the consumer price indexes, the monthly changes in the prices paid by urban consumers for a representative basket of goods and services of all items (not seasonally adjusted). Data source: Bureau of Labor Statistics,
U.S. Department of Labor (http://www.bls.gov/cpi/).
2. Industrial production index, INDPRO: we use the monthly changes of real output for all facilities located in the United States manufacturing, mining, and electric and gas utilities (excluding those in U.S. territories). Data source: Board of Governors of the Federal Reserve System (http://www.federalreserve.gov/).
3. Stock market index: we use one-month changes of the S&P 500 index where the Standard & Poor's 500 (S&P 500) is an American stock market index based on the market capitalization of 500 large companies having common stock listed on the New York stock exchange (NYSE) or the National Association of Securities Dealers Automated Quotations (NASDAQ). Data source: Yahoo Finance (https://finance.yahoo.com).
4. Market volatility: we use the Chicago Board Options Exchange Market (CBOE) Volatility Index, VIX. Data source:
Chicago Board of Options Exchange (http://www.cboe.com/).

We next elect five stock return factors from two groups: the valuation group and the growth group. We collect stock characteristic (factor) data for all stocks in the S&P 500 from January 1990 to December 2014. Stock price data was provided by Mergent and stock fundamental data (e.g., earnings, sales) were provided by Compustat. In the stock valuation group, we named the three following factors:

1. Earnings/price (E/P) is calculated by the earning accumulation over the trailing twelve months of the stock divided by weekend price. A higher number indicates greater value for each unit of earnings, which tends to drive higher stock returns.

2. The free cash flow/enterprise value is calculated by the cash flow minus cash dividends minus capital expenditures divided by market value of equity plus debt. A higher number is better.

3. The sales/enterprise value is calculated by the sales accumulation over the trailing twelve months of the stock divided by the market value of equity plus debt (enterprise value). A higher sales/enterprise value signifies that each unit of a stock's value is used to generate more sales, which normally leads to higher stock returns.

All valuation measures are presented as yields (e.g., earning/price instead of the traditional form price/earning) for more accurate, comparative purposes. Raw fundamental data can provide values close or equal to zero, which heavily skew traditional valuation metrics. Inverting the traditional valuation calculation (i.e., creating valuation yield measures) mitigates this problem and also allows for direct comparison to non-equity investments (e.g., bond yields). Furthermore, valuation yields better accommodate the display of distributions containing negative numbers (more robust summary statistics).

In the stock growth group, we choose the following two factors:

1. Long-term earning per share (LT EPS) growth: projected long-term growth rate of earning per share based on a five-year moving regression trend line. A high earnings growth rate normally leads to higher future returns.

2. Long-term sales (LT sales) growth: projected long-term growth rate of sales based on a five-year moving regression trend line. A high sales growth rate normally leads to higher future returns.

As growth factors, they can be used in either a trend-following or a contrary way depending on the universe and macro environment.

4.4.3 Model Description

Using these above variables, our stock selections were divided into two steps. The first step is to find regimes of macro variables and the second step is to make stock selection. Among these four macroeconomic indicators, CPI and INDPRO are monthly data while VIX and the S&P 500 index are daily data. Thus, we use monthly frequency to take advantage of fresh data. Furthermore, using monthly data will give a moderate rotation rate, or the average of time (months) a stock stays in the portfolio, for our composite portfolio.

First, we calibrate the HMM parameters, $\lambda = (\mathbf{A}, \mu, \sigma, \mathbf{p})$, using the Baum-Welch algorithm, (from Section 2.5), and one of the four macroeconomic variables above. We then use the obtained parameters to predict the corresponding hidden regimes of each economic indicator using the Viterbi algorithm (from Section 2.4). We use monthly historical data of the variables from January 1990 to December 2014 for the first step. In the second step, we make monthly selections starting from December 1999 to December 2014. In this step, we choose fifty stocks based on their characteristic (factor) performances. Each month, after predicting economic regimes of the four macroeconomic variables (CPI, INDPRO, S&P 500, and VIX) for a period from the target month back to the past (January 1990) in Step 1, we look back in history for periods with similar regimes as those of the next month and examine the five stock characteristics (defined in Section 4.7.4) to determine how well the stock factors performed in these periods. We then give score and weight to each factor and form a composite score for each stock of the S&P 500. Finally, we select a portfolio of the best 50 stocks (top decile) and compare the performances of the portfolio over time with the benchmark index (S&P 500). We continue the stock evaluation process by updating

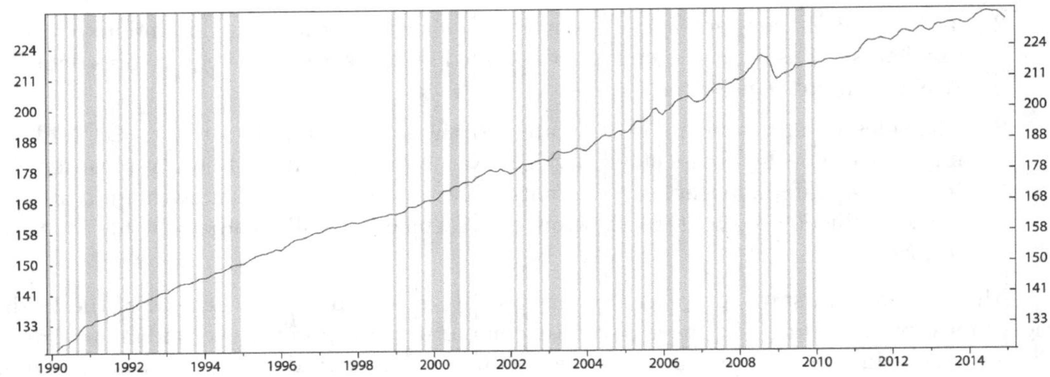

Regime Parameters (2014-12-31)

Mu		Sigma	
Inflation (Unshaded)	Deflation (Shaded)	Inflation (Unshaded)	Deflation (Shaded)
2.80	7.28	0.93	3.55

Data Statistics

Mean		Variance	
Inflation (Unshaded)	Deflation (Shaded)	Inflation (Unshaded)	Deflation (Shaded)
2.37	2.89	0.69	2.26

FIGURE 4.15
Regimes of inflation, CPI; monthly data from January 1990 to December 2014 (log scale).

the parameters of the HMM and adding the value of the most recent month for the macro variables, to make a new prediction. We will present more details about the processes in the next sections.

4.4.4 Identifying Regimes in Macroeconomic Variables

Our first approach to stock selection is using the HMM to find regimes for four economic macro variables: inflation, industrial production index, stock market index, and market volatility. We focus on finding only two opposite states of these four economic indicators to keep the model simple while maintaining the predictive power as reasonable. In addition, most macro variables often experience two opposite states: Bull/Bear for the stock index, S&P 500, inflation/deflation for inflation, CPI, low/high volatility for market volatility, VIX, and growth/recession for the industrial production index, INDPRO. Therefore, we just need two states, State 1 and State 2, for our model. We define State (or Regime) 2 as the regime corresponding to the normal distribution that has the lower ratio of mean and variance. For the stock market variable, State 1 represents the bull market, and State 2 represents the bear market. For the inflation variable, State 1 stands for inflation and State 2 stands for deflation. For the industrial production index variable, State 1 and State 2 interpret growth and recession, respectively. For the market volatility, State 1 represents low volatility and State 2 represents high volatility.

We use historical data of each variable and record new data monthly from January 1990 to December 2014 to calibrate HMM parameters and to find the corresponding regimes. The results are shown in Figures 4.15-4.18.

On Figures 4.15-4.18, the unshaded areas represent State 1 and the shaded areas represent State 2. Under each of these figures are two tables: in one table, we report the calibrated parameters of the HMM. In the other table, we summarize the mean and variance of each macro variable on the corresponding regime. In all four figures, the performance of the economic indicators on each regime (in terms of center and spread) is consistent with the

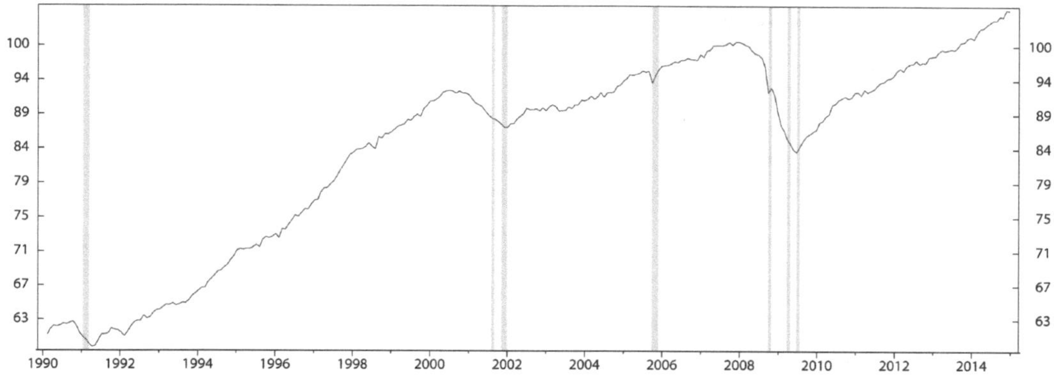

Regime Parameters (2014-12-31)

Mu		Sigma	
Growth (Unshaded)	Recession (Shaded)	Growth (Unshaded)	Recession (Shaded)
0.31	-0.47	0.51	1.27

Data Statistics

Mean		Variance	
Growth (Unshaded)	Recession (Shaded)	Growth (Unshaded)	Recession (Shaded)
0.21	-0.57	0.24	1.82

FIGURE 4.16
Regimes of the industrial production index, INDPRO; monthly data from January 1990 to December 2014 (log scale).

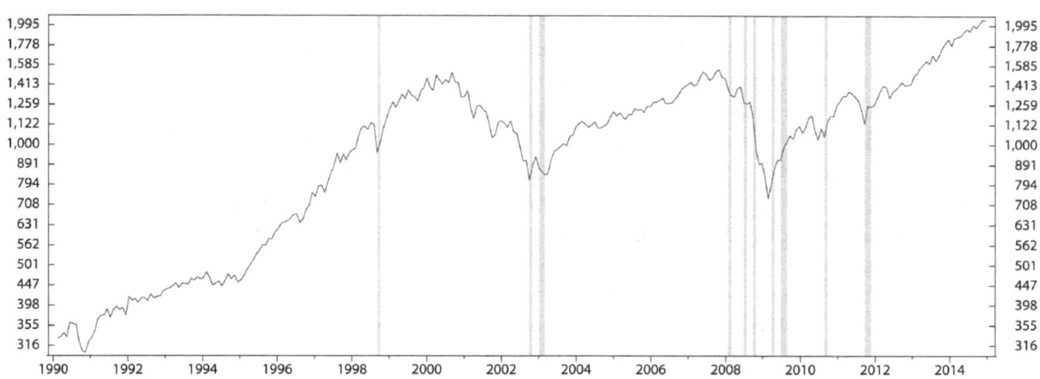

Regime Parameters (2014-12-31)

Mu		Sigma	
Bull (Unshaded)	Bear (Shaded)	Bull (Unshaded)	Bear (Shaded)
0.57	0.97	0.51	1.27

Data Statistics

Mean		Variance	
Bull (Unshaded)	Bear (Shaded)	Bull (Unshaded)	Bear (Shaded)
1.02	-0.76	12.30	59.48

FIGURE 4.17
Regimes of the stock market index, S&P 500; monthly data from January 1990 to December 2014 (log scale).

definition of regimes. Figure 4.15 and Figure 4.17 show the regimes of inflation and the stock market index, respectively. The inflation rate, CPI, has a slightly higher mean during the deflation periods; however, it also has higher variance compared to those during the inflation periods. The stock market index, S&P 500, has a negative average return and

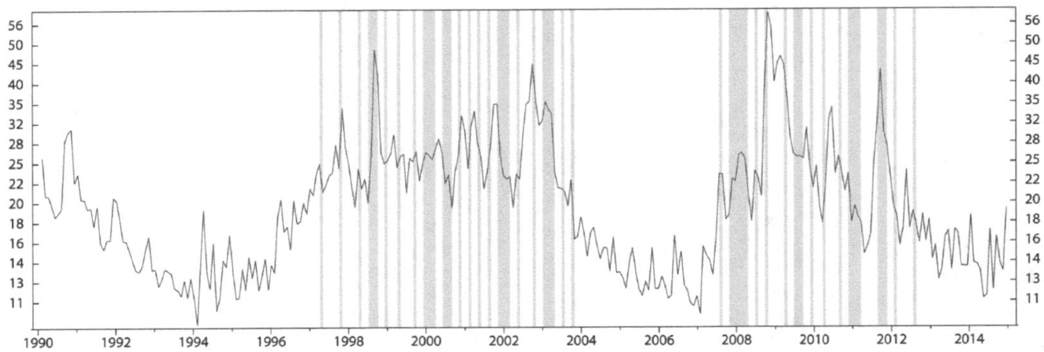

Regime Parameters (2014-12-31)

Mu		Sigma	
Low Volatility (Unshaded)	High Volatility (Shaded)	Low Volatility (Unshaded)	High Volatility (Shaded)
14.80	25.92	2.50	7.62

Data Statistics

Mean		Variance	
Low Volatility (Unshaded)	High Volatility (Shaded)	Low Volatility (Unshaded)	High Volatility (Shaded)
15.47	26.75	12.07	54.44

FIGURE 4.18
Regimes of market volatility, VIX; monthly data from January 1990 to December 2014 (log scale).

high volatility periods during the Bear market, while it performs well in the Bull market. Figure 4.16 shows the regimes of the industrial production index, INDPRO. The INDPRO has a lower mean and higher variance of growth rate during recessions than those during the growths. Figure 4.18 shows the regimes of the market volatility, VIX. We find that VIX is more stable during high volatility periods than during low volatility periods.

The following Figures 4.19-4.21 represent performances of the S&P 500 index during regimes of industrial production index, market volatility, and inflation from January 1950 to July 2015 respectively. The results above indicate that using different economic indicators the HMM gives different predictions for economic regimes. Furthermore, it is clear from the seven Figures 4.15-4.21 that each of the macro variables was in State 2 during the economic crisis period from 2008 to 2010. We have the conclusion that the HMM can predict economic crisis by using a proper economic indicator.

4.4.5 Stock Selection

The motivation behind this stock selection idea originated from observations that the stock market performs significantly different across different regimes, Figures 4.19-4.21. Our question is: can we use the performances of stocks in the past to predict their future performances? An index seems to have similar behaviors on the same economic regimes. Thus, it is reasonable to analyze stock characteristics (or stock factors) from a similar environment in the past to forecast its future outcomes. Based on this motivation, in this section, we discuss how to use the HMM to make stock selections.

We investigate the stock factors that were affected the most by the macro regimes and use these factors to rank stocks in the S&P 500. Each month, we select 50 stocks from the S&P 500 universe based on the stock ranking to add into our portfolio. At the end of each month, we first calibrate the HMM parameters using monthly data of each of the four macroeconomic variables: inflation, industrial production index, market index, and market

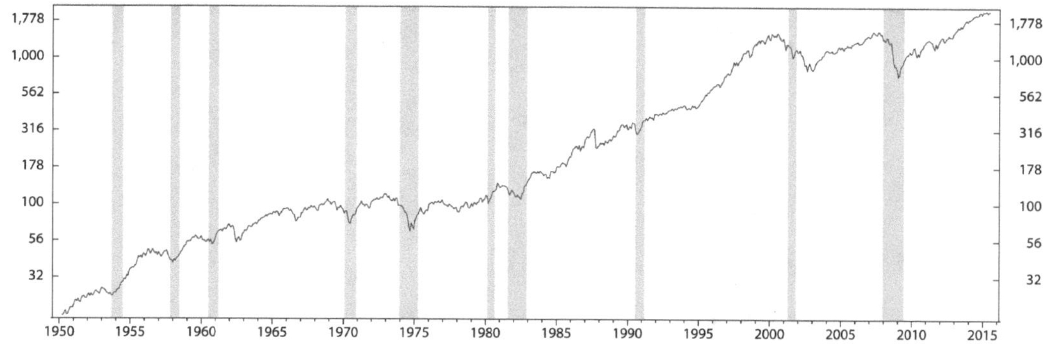

Regime Statistics

Annualized Returns		Annualized Standard Deviation	
Normal (Unshaded)	Recession (Shaded)	Normal (Unshaded)	Recession (Shaded)
7.80	6.49	13.37	18.60

FIGURE 4.19
S&P 500 performances vs. the industrial production index regimes; monthly data January 1950 to July 2015 (log scale).

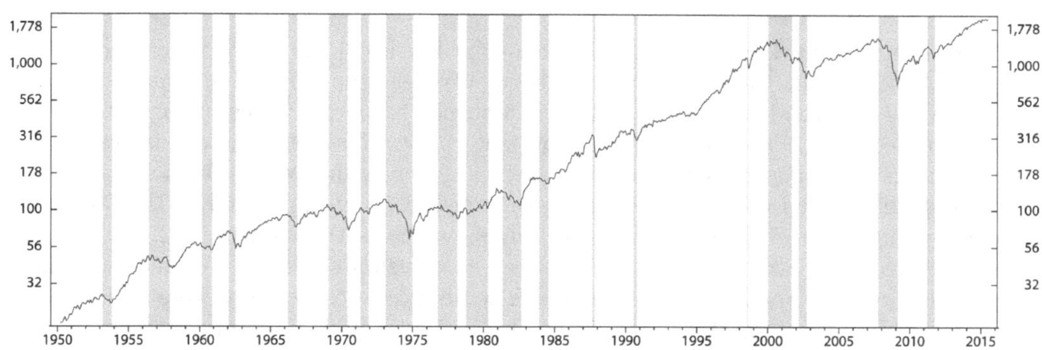

Regime Statistics

Annualized Returns		Annualized Standard Deviation	
Bull (Unshaded)	Bear (Shaded)	Bull (Unshaded)	Bear (Shaded)
18.95	-15.38	11.50	17.91

FIGURE 4.20
S&P 500 performances vs. the market volatility's regimes; monthly data January 1950 to July 2015 (log scale).

volatility. We then find the regimes of each variable during the time period and predict the upcoming regimes. After predicting the regimes of the four variables in the next month, we look back at historical data (twenty years) and find similar performances of these four variables. For example, if the predicted regimes for the next month of inflation, industrial

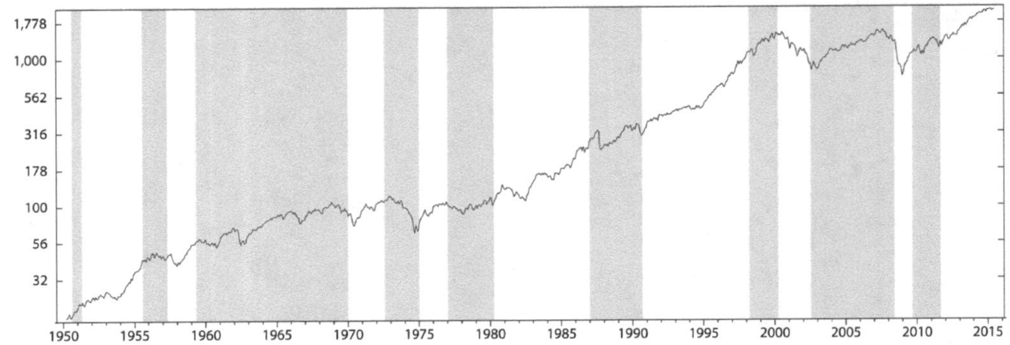

Regime Statistics

Annualized Returns		Annualized Standard Deviation	
Normal (Unshaded)	Inflation (Shaded)	Normal (Unshaded)	Inflation (Shaded)
11.29	4.19	13.78	14.70

FIGURE 4.21
S&P 500 performances vs. the inflation regimes; monthly data January 1950 to July 2015 (log scale).

production index, market index, and market volatility are 1, 2, 2, and 1, respectively, we will then look back at the data for the past twenty years and find the months that these four variables had the same regimes. We then check the performance of each stock factor (defined in Section 4.7.4: E/P, free cash flow/enterprise value, sales/enterprise value, long-term earnings per share growth, and long-term sales growth). We rank each stock factor from one to five based on its performance, then assign its weight equal to the factor rank divided by the sum of all of the possible ranks (which is the sum of 1, 2, 3, 4, and 5). The composite score for each stock is calculated by the summation of products of its factor ranks with the corresponding factor weights. The final composite scores were scaled to have the range from 1 to 100. We select 50 stocks with the highest composite score for our portfolio. We sell stocks that are not in the election list while buying the newly-entered ones. Our calculation shows that once a stock is added to our portfolio, it stays for about three months. Figures 4.22-4.26 show the monthly performances of the stock factors of the 50 selected stocks. Each of the figures consists of two graphs. The graph on the top compares the consummation of our portfolio's factor vs. the S&P 500 returns from December 1999 to December 2014. The graph on the bottom presents the excess return and the accumulated return of the factors of our 50 chosen stocks. We scale the stock prices so that the price at the beginning of the stock electing process (December 1999) could be \$100.00. At the bottom of each figure is a table that summarizes the performances. We calculate the annual gain (gain/annum), standard deviation, downside deviation, batting average, Sharpe ratio, information ratio (info ratio), tracking error, and maximum draw down of the data. The batting average is a statistical measure used to measure an investment model's ability to meet or beat an index. The batting average is calculated by dividing the number of days (or months, quarters, etc.) in which the model beats or matches the index by the total number of days (or months, quarters, etc.) in the period and then multiplying that factor by 100. In this book, the batting average is the percent of time the portfolio returns meet or exceed the S&P 500 index returns. The tracking error measures the consistency of excess

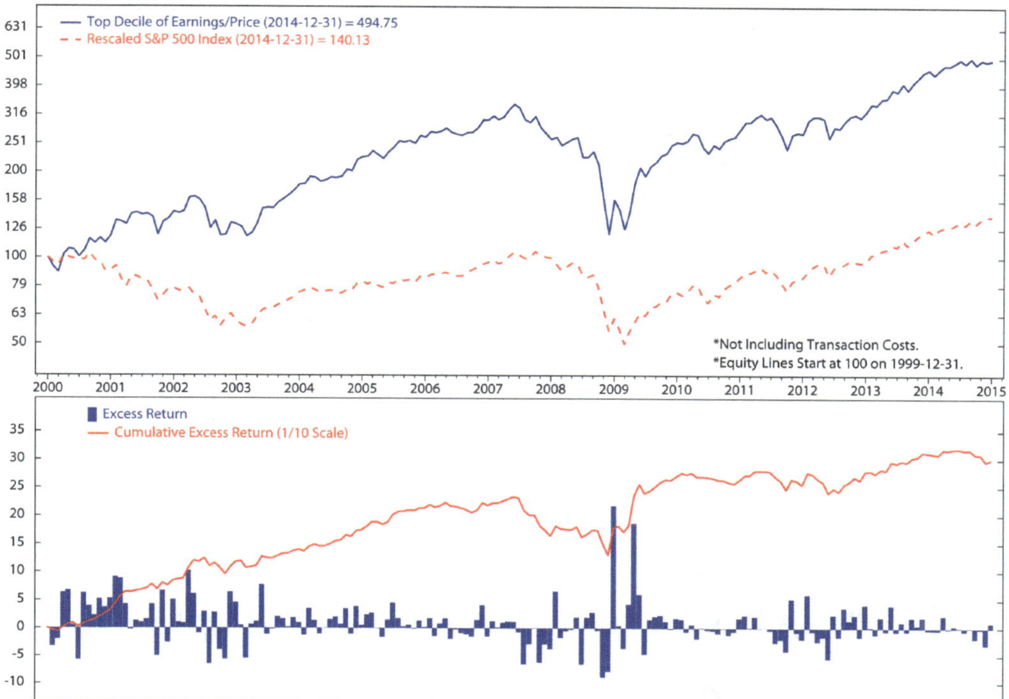

Title	Gain/ Annum	Standard Deviation	Downside Deviation	Batting Average	Sharpe Ratio	Info Ratio	Tracking Error	Max Drawdown
Top Decile of Earnings/Price	11.2%	24.1%	18.0%	58.3%	0.39	0.69	13.0%	-64.9% (2007-05-31..2008-12-01)
S&P 500 Index	2.3%	15.7%	12.3%		0.03			-52.5% (2007-09-30..2009-03-01)

FIGURE 4.22
Comparison of earnings/price and the S&P 500; monthly data from December 1999 to December 2014 (log scale).

returns. It is created by taking the difference between the model return and the benchmark return every month or quarter and then calculating how volatile that difference is. Tracking error is also useful in determining just how "active" a model's strategy is. The lower the tracking error, the closer the model follows the benchmark. The higher the tracking error, the more the model deviates from the benchmark. The Sharpe ratio (also known as the Sharpe index, the Sharpe measure, and the reward-to-variability ratio) is a way to examine the performance of an investment by adjusting for its risk. The ratio measures the excess return (or risk premium) per unit of deviation in an investment asset or a trading strategy, typically referred to as risk (and is a deviation risk measure). In this book, the Sharpe ratio is calculated by taking the ratio of excess returns vs. S&P 500 index and its standard deviation.

We make an assumption that at the end of each month, we would sell all of the stocks in our portfolio that are not on the list of the new 50 elected stocks and buy the new named stocks for the next month without transaction fees. The transaction cost is expected to be minimal in our model, because, based on our portfolio performances, the holding period of the portfolio constituents is about three months. Figures 4.22-4.26 show clearly that the stock returns and the valuation factors dropped significantly during the economic crisis

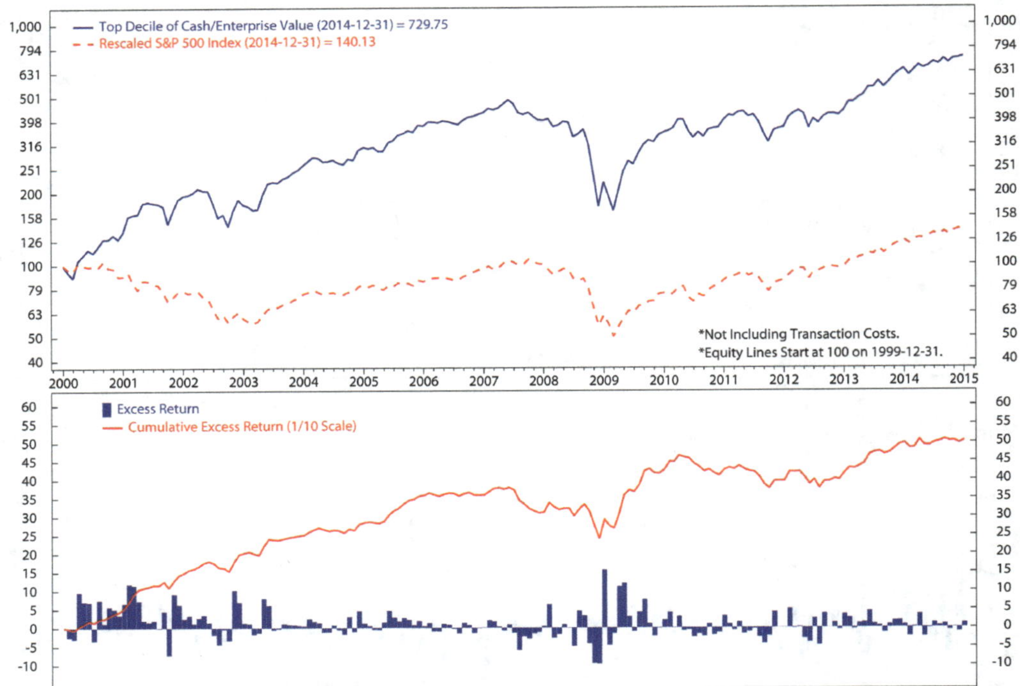

Title	Gain/ Annum	Standard Deviation	Downside Deviation	Batting Average	Sharpe Ratio	Info Ratio	Tracking Error	Max Drawdown
Top Decile of Cash/Enterprise Value	14.2%	24.5%	19.2%	59.4%	0.50	0.93	12.8%	-65.7% (2007-05-31..2009-03-01)
S&P 500 Index	2.3%	15.7%	12.3%		0.03			-52.5% (2007-09-30..2009-03-01)

FIGURE 4.23
Comparison of free cash flow/enterprise value and the S&P 500; monthly data from December 1999 to December 2014 (log scale).

starting from the third quarter of 2007 to the first quarter of 2009. The five factors of the preferred stocks perform better than the S&P 500 in terms of the accumulated return.

Stock factor returns are significantly different in each economic regime. Thus, we assign the weight for each factor based on its execution. Figure 4.27 represents the weights of each stock factor monthly from December 1999 to December 2014.

Figure 4.28 shows the accumulated returns of our stock selection portfolios vs. that of the S&P 500 returns. Our stock portfolio has higher returns than the S&P 500. Based on our model, with an initial investment of $100, in the 15 years from December 1999 through December 2014, our portfolio had an average gain per annum of 14.9% vs. 2.3% for the S&P 500. The gains were calculated without transaction fees. Our portfolio returns were the result of monthly trading; on the other hand, the earnings of the S&P 500 were estimated by buy and hold for the long term, which is for fifteen years. Our portfolio performance is also better than the performance of any single factor strategy. Compared to the best factor strategy of long top 50 stocks of the free cash flow/enterprise value during the period, Figure 4.23, our portfolio shows an excess return of 0.7% (i.e., 14.9% vs. 14.2%). Most importantly, our portfolio also has lower risk, thanks to the diversification of risk across various factor strategies. As a result, its Sharpe ratio (i.e., return/risk) is higher than that of the free cash flow/enterprise value factor strategy (0.61 vs. 0.5).

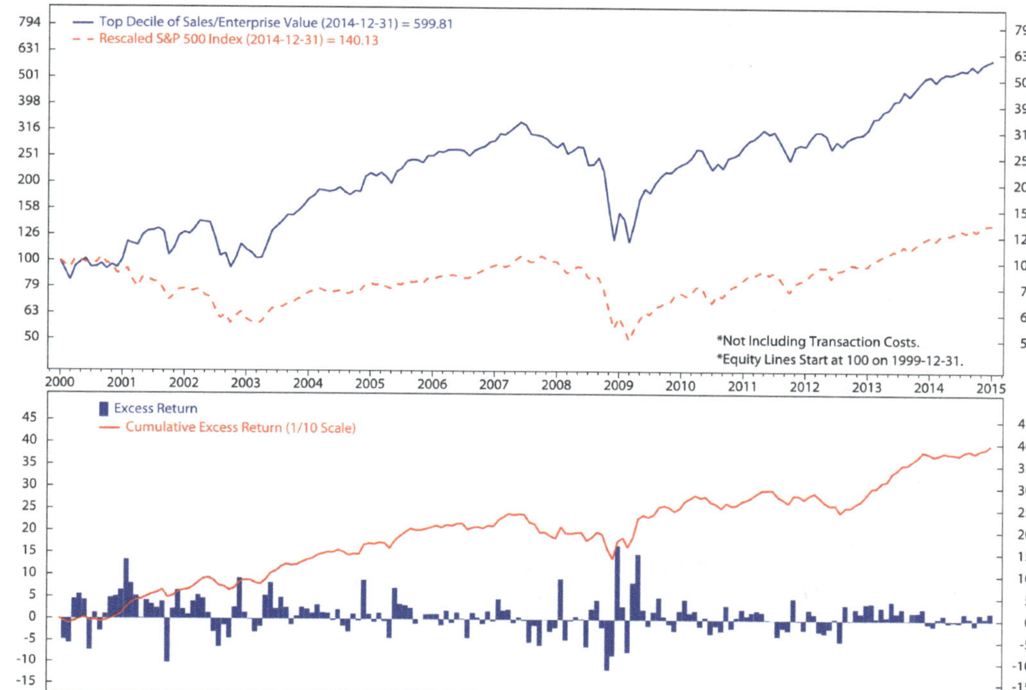

Title	Gain/ Annum	Standard Deviation	Downside Deviation	Batting Average	Sharpe Ratio	Info Ratio	Tracking Error	Max Drawdown
Top Decile of Sales/Enterprise Value	12.7%	24.8%	19.8%	63.9%	0.44	0.78	13.4%	-65.1% (2007-05-31..2009-03-01)
S&P 500 Index	2.3%	15.7%	12.3%		0.03			-52.5% (2007-09-30..2009-03-01)

FIGURE 4.24
Comparison of sales/enterprise value and the S&P 500; monthly data from December 1999 to December 2014 (log scale).

4.4.6 Conclusions

Inflation, the industrial production index, the market index and the market volatility have significant impacts on the stock market. Stock performances differ during regimes of those macroeconomic indicators. Each month, we used the HMM to predict the regimes for all of these variables for the upcoming month and tracked the historical data to find similar macro environments. After finding the similar periods, we examined five stock characteristics to determine their scores and weights based on their rewarded returns. We assigned the composite score for each stock in the S&P 500 universe based on the scores and weights of its factors. The weights of the five stock factors are presented in Figure 4.27. Finally, we chose the 50 top ranked stocks and held for at least a month. Our back tests showed that using the HMM for stock selections made significant portfolio returns compared to those of the S&P 500 universe. The results are shown in Figures 4.28.

4.5 HMM for Stock Trading

The HMM is a statistical signal prediction model which has been widely used to predict economic regimes and stock prices. In this study, we introduce the application of the HMM

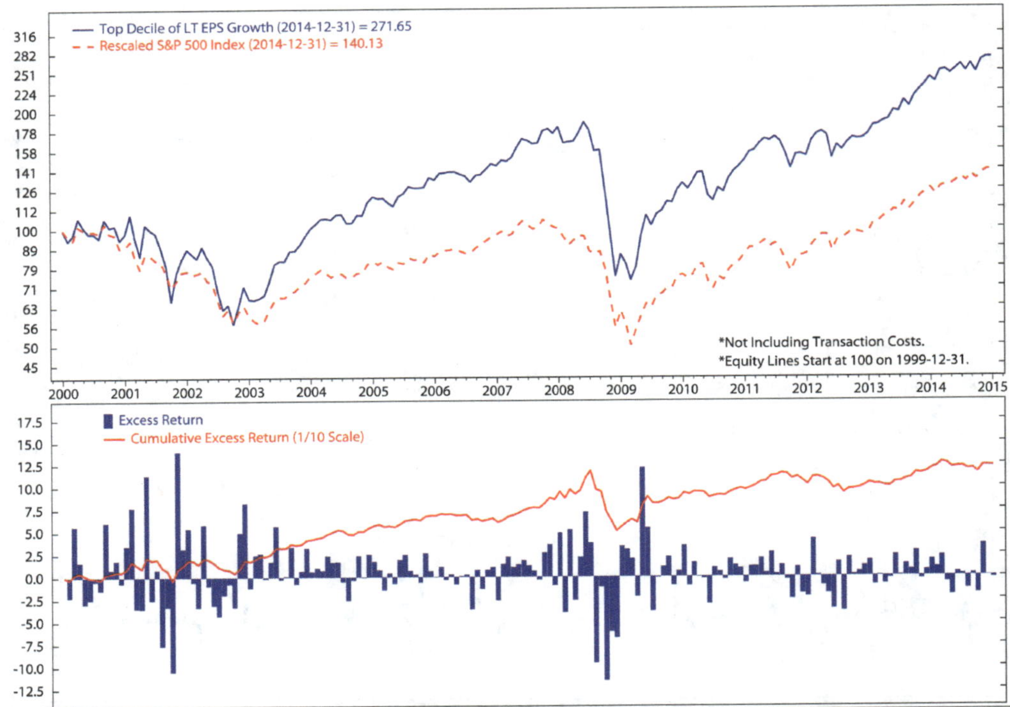

Title	Gain/ Annum	Standard Deviation	Downside Deviation	Batting Average	Sharpe Ratio	Info Ratio	Tracking Error	Max Drawdown
Top Decile of LT EPS Growth	6.9%	23.4%	18.8%	59.4%	0.22	0.41	11.4%	-60.6% (2008-05-31..2009-03-01)
S&P 500 Index	2.3%	15.7%	12.3%		0.03			-52.5% (2007-09-30..2009-03-01)

FIGURE 4.25
Comparison of top decile of long-term earning per share (LT EPS) growth vs. the S&P 500; monthly data from December 1999 to December 2014 (log scale).

in trading stocks (with the S&P 500 index being an example) based on the stock price predictions. The procedure starts by using four criteria, including the Akaike information, AIC [2], the Bayesian information, BIC [124], the Hannan Quinn information, HQC [60], and the Bozdogan Consistent Akaike Information, CAIC [19], in order to determine an optimal number of states for the HMM. The selected four-state HMM is then used to predict monthly closing prices of the S&P 500 index. For this work, the out-of-sample R^2_{OS}, and some other error estimators are used to test the HMM predictions against the HAM. Finally, both the HMM and the HAM are used to trade the S&P 500. The obtained results clearly prove that the HMM outperforms this traditional method in predicting and trading stocks.

4.5.1 Introduction

Stock traders always wish to buy a stock at a low price and sell it at a higher price. However, when is the best time to buy or sell a stock is a challenging question. Stock investments can have a huge return or a significant loss due to the high volatility of the stock prices. Many models were used to predict stock executions such as the "exponential moving average" (EMA) and the "head and shoulders" methods. However, many of the forecast models require a stationary of the input time series. In reality, financial time series are often

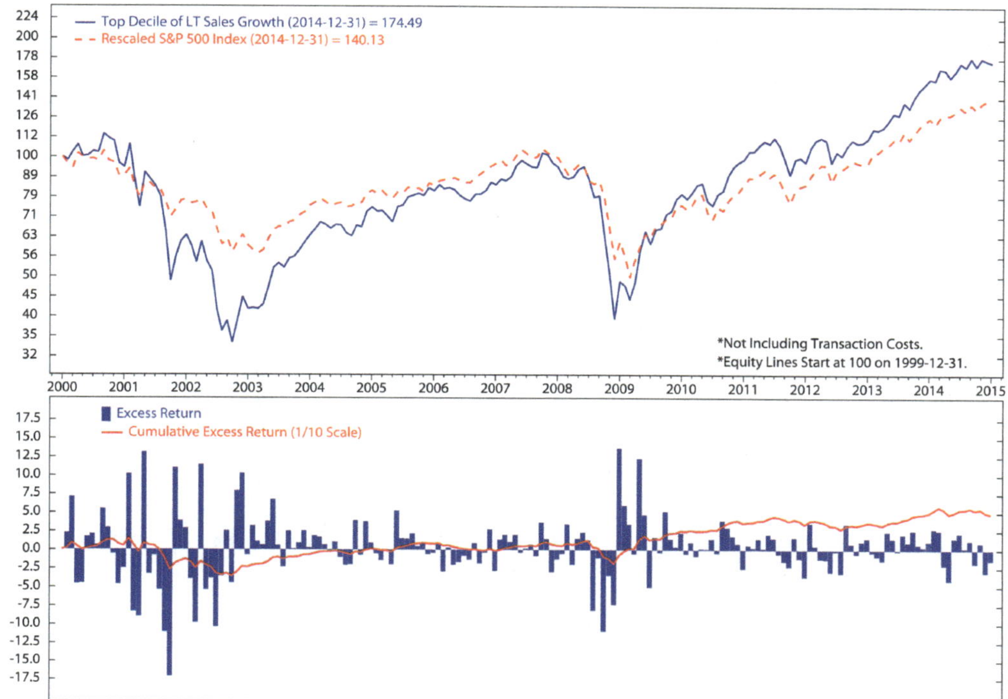

Title	Gain/ Annum	Standard Deviation	Downside Deviation	Batting Average	Sharpe Ratio	Info Ratio	Tracking Error	Max Drawdown
Top Decile of LT Sales Growth	3.8%	25.9%	20.3%	57.2%	0.08	0.11	14.2%	-70.1% (2000-08-31..2002-10-01)
S&P 500 Index	2.3%	15.7%	12.3%		0.03			-52.5% (2007-09-30..2009-03-01)

FIGURE 4.26
Comparison of top decile of LT sales growth vs. the S&P 500; monthly data from December 1999 to December 2014 (log scale).

non-stationary, thus nonstationary time series models are needed. Autoregression models have been modified by adding time-dependent variables to adapt with the nonstationary of time series. The author of [54] developed the threshold auto-regressive model where its parameters depend on the value of a previous observation. The researchers of [48, 73, 148] introduced mixture models for time series. These models are the special case of the Markov Switching auto-regression models.

Based on the principal concepts of the HMM, the author of [59] developed a regime-switching model (RSM) for a nonlinear stationary time series and business cycles. Researchers, such as those cited in [45,64,123,127], have shown that regime-switching variables significantly improve portfolio selection models. However, the RSM is not identical to the HMM introduced in [12]. The RSM's parameters were generated by an auto-regression model while the HMM parameters were calibrated by maximizing the log-likelihood of observation data. Although the RSM and the HMM are both associated with regimes, or hidden states, the RSM should be viewed as a regression model that has a regime-shift variable as an explanatory variable. Furthermore, the HMM is a broader model that allows a more flexible relationship between observation data and its hidden state sequence. This relationship can be presented by the observation probability functions corresponding to each hidden state. The common functions that researchers used in the HMM models are the density function of normal, mixed normal, or exponential.

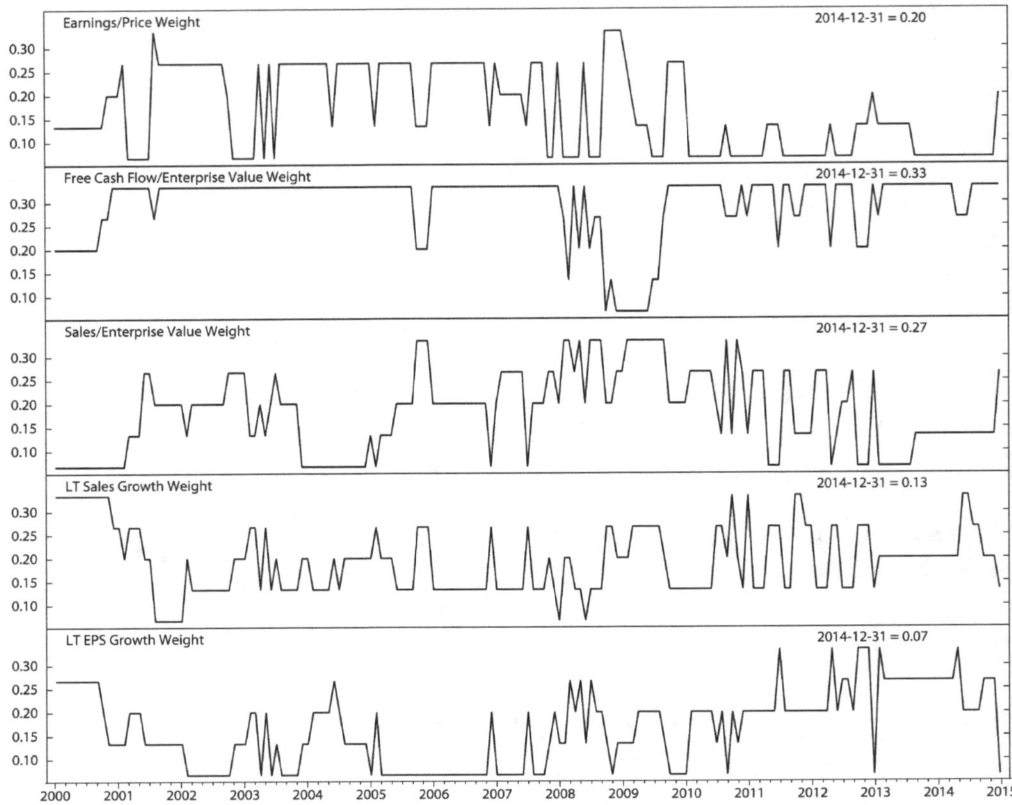

FIGURE 4.27
Weight of five stock factors; monthly data from December 1999 to December 2014.

Recently, researchers have applied the HMM to forecast stock prices. The author of [61] used the HMM to predict the stock price for inter-related markets. The HMM with two states was used to predict regimes in market turbulence, inflation, and industrial production index [79]. Furthermore, the HMM with four states was used with multiple observations to study asset allocation decisions based on regime switching in asset returns [58]. Similarly, the authors of [5] applied the RSM for international asset allocation. The author of [103] used the HMM with both single and multiple observations to forecast economic regimes and stock prices. The HMM with multiple observation data (open, close, low, and high prices of a stock) was used to predict its closing prices [108]. In our previous work [105], we used the HMM for a single observation sequence to predict the regimes of some economic indicators and to make stock selections based on the performances of these stocks during the predicted schemes.

In this study, we use the HMM for multiple independent observation sequences: the "open", "low", "high", and "close" prices of the U.S. benchmark index S&P 500. We limit the number of states of the HMM at six to keep the model simple and feasible with cyclical economic stages. Four criteria used to select the best HMM model are:

1. the Akaike information criterion (AIC) [2]
2. the Bayesian information criterion (BIC) [124]

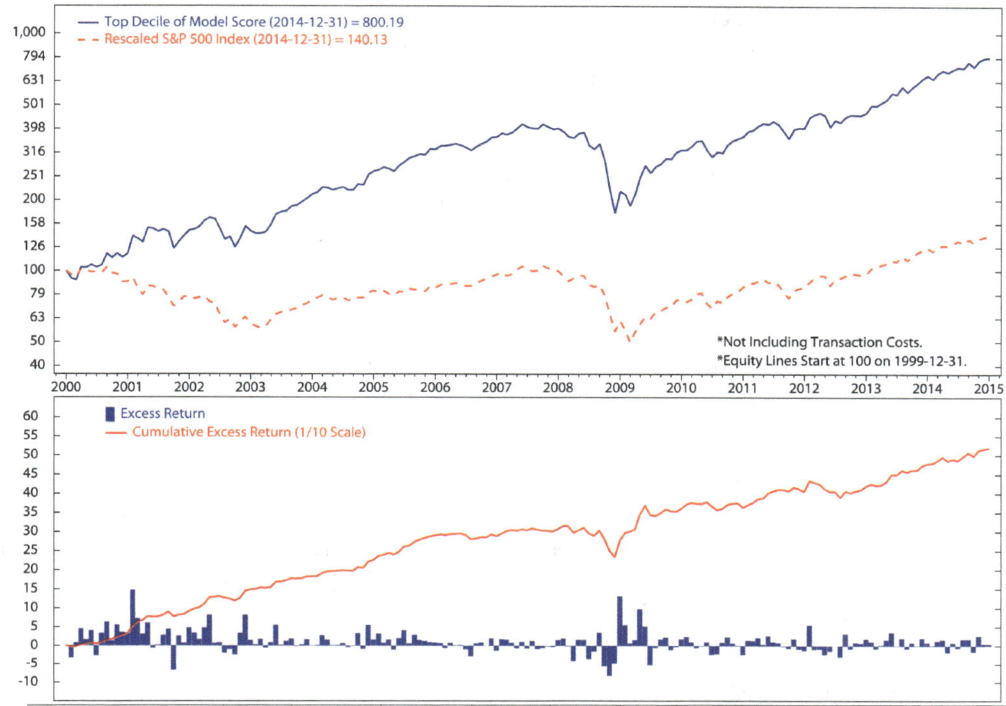

Title	Gain/ Annum	Standard Deviation	Downside Deviation	Batting Average	Sharpe Ratio	Info Ratio	Tracking Error	Max Drawdown
Top Decile of Model Score	14.9%	21.2%	16.7%	68.3%	0.61	1.27	9.9%	-57.9% (2007-05-31..2008-12-01)
S&P 500 Index	2.3%	15.7%	12.3%		0.03			-52.5% (2007-09-30..2009-03-01)

FIGURE 4.28
Comparison of stock selected portfolio using the HMM and the S&P 500 from December 1999 to December 2014.

3. the Hannan-Quinn information criterion (HQC) [60]
4. the Bozdogan Consistent Akaike Information Criterion (CAIC) [19].

After selecting the best model, we use the HMM to predict the S&P 500 price and compare the results with that of the historical average return model (HAR). Finally, we apply the HMM and the HAR models to trade the stock and confront their results.

The stock price prediction process is based on the work of [61]. The authors used the HMM with the four observations: close, open, high, and low price of some airline stocks to predict their future closing price using four states. In their work, the authors find a day in the past that was similar to the recent day and used the price change in that date and price of the recent day to predict a future closing price. Our approach is different from their work in three modifications. First, we use the AIC, BIC, HQC, and the CAIC to test the performances of the HMM with two to six states. Second, we use the selected HMM model and multiple observations (open, close, high, and low prices) to predict the closing price of the S&P 500. Many statistical methods are used to evaluate the HMM out-of-sample predictions over the results obtained by the benchmark HAR model. Third, we also go further by using the models to trade the S&P 500 using different training and predicting periods.

TABLE 4.5
Summary of S&P 500 index monthly prices from 01/03/1950 to 11/01/2016

Price	Min	Max	Mean	Std.
Open	16.66	2,173.15	504.06	582.20
High	17.09	2,213.35	520.46	599.28
Low	16.66	2,147.56	486.91	562.96
Close	17.05	2,213.35	506.73	584.98

To test our model for out-of-sample predictions, we use the out-of-sample R^2, [23], and the cumulative squared predictive errors, [163] to compare the performances of the HMM and the HAM.

4.5.2 Data Selection

We choose one of the common benchmarks for the U.S. stock market, the S& P 500 index to implement our model. The S&P 500 was recorded as monthly data from January 1950 to November 2016 and was taken from Yahoo Finance (https://finance.yahoo.com). The summary of the data is presented in Table 4.5.

4.5.3 Model Selection

Choosing a number of hidden states for an HMM is a critical task. In the section, we use four common criteria: the AIC, the BIC, the HQC, and the CAIC that were introduced in Section 3.1 to evaluate the performances of the HMM with different numbers of states. These criteria are suitable for the HMM because in the model training algorithm, i.e. the Baum-Welch algorithm, the EM method was used to maximize the log-likelihood of the model. We limit numbers of states from two to six to keep the model simple and feasible to stock prediction. Recall that these criteria are calculated using the following formulas respectively:

$$\text{AIC} = -2\ln(L) + 2k \tag{4.20}$$

$$\text{BIC} = -2\ln(L) + k\ln(M) \tag{4.21}$$

$$\text{HQC} = -2\ln(L) + k\ln(\ln(M)) \tag{4.22}$$

$$\text{CAIC} = -2\ln(L) + k(\ln(M) + 1), \tag{4.23}$$

where L is the likelihood function for the model, M is the number of observation points, and k is the number of estimated parameters in the model. In this book, we assume that the distribution corresponding with each hidden state is a Gaussian distribution, therefore the number of parameters, k, is formulated as $k = N^2 + 2N - 1$, where N is numbers of states used in the HMM.

To train the HMM parameters, we use a historical observed data of fixed length T,

$$O = \{O_t^{(1)}, O_t^{(2)}, O_t^{(3)}, O_t^{(4)},\ t = 1, 2, ..., T\},$$

where $O^{(i)}$ with $i = 1, 2, 3,$ or 4 represents the "open", "low", "high" or "close" price of a stock respectively. For the HMM with a single observation sequence, we use only closing price data,

$$O = O_t,\ t = 1, 2, ..., T,$$

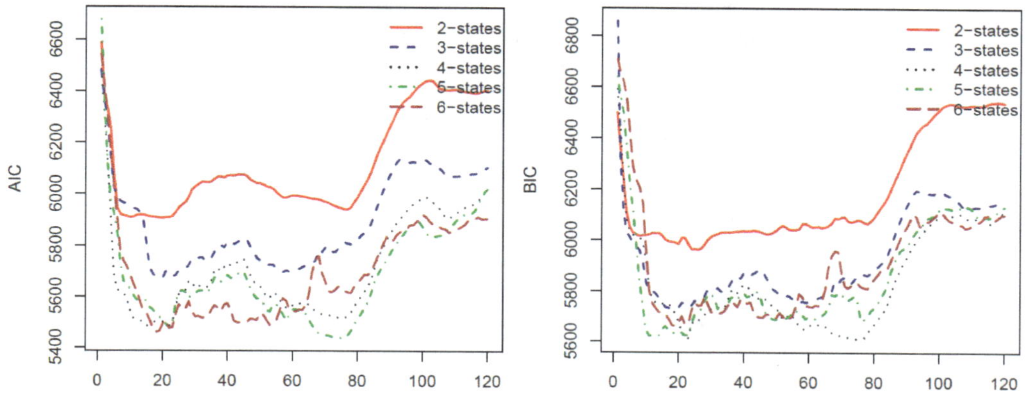

FIGURE 4.29
AIC (left) and BIC (right) for 120 HMM parameter calibrations using S&P 500 monthly prices.

where O_t is the stock closing price at time t. We use S&P 500 monthly data to train the HMM and calculate the criteria AIC, BIC, HQC, and CAIC. Each time, we use a block of length $T = 120$ (ten year period) of S&P 500 prices, $O = \{\text{"}open\text{"}, \text{"}low\text{"}, \text{"}high\text{"}, \text{"}close\text{"}\}$, to calibrate the HMM parameters and calculate the AIC and BIC numbers. The first block of data is the S&P 500 monthly prices from December 1996 to November 2006. We use the set of data to calibrate the HMM parameters, using the parameters to calculate the probability of the observations, which is the likelihood L of the model. Then, we use the likelihood to calculate the criteria using formulas 4.20, 4.21, 4.22, and 4.23. We choose initial parameters for the first calibration as follows:

$$\begin{aligned} A =&(a_{ij}),\ a_{ij} = \frac{1}{N}, \\ p =& \left[1, 0, .., 0\right], \\ \mu_i =& \mu^{(O)} + Z,\ Z \sim \mathcal{N}(0,1), \\ \sigma_i =& \sigma^{(O)}, \end{aligned} \tag{4.24}$$

where $i,\ j = 1, .., N$ and $\mathcal{N}(0, 1)$ is the standard normal distribution. For the second calibration, we move the ten-year data upward one month giving us a new data set from January 1997 to December 2006 with which we can use the calibrated parameters from the first calibration as initial parameters. We repeat the process 120 times for 120 blocks of data by moving the block of data forward. The last block of data is the monthly prices from November 2006 to November 2016. The AIC, BIC, HQC, and CAIC of the 120 calibrations are presented in Figures 4.29-4.30. In all of these four criteria, a model with a lower criterion value is better. Thus, the results from Figure 4.29 and Figure 4.30 show that the four-state HMM is the best model among the two to six state HMMs. Therefore, we will use the HMM with four states in stock price predicting and trading. We want to note that for a different stock, using these criteria we may have another optimal number of states for the HMM. Therefore, we suggest that before applying the HMM to predict prices for a stock, researchers should use some of the presented criteria to choose a number of states of the HMM that works the best for the stock.

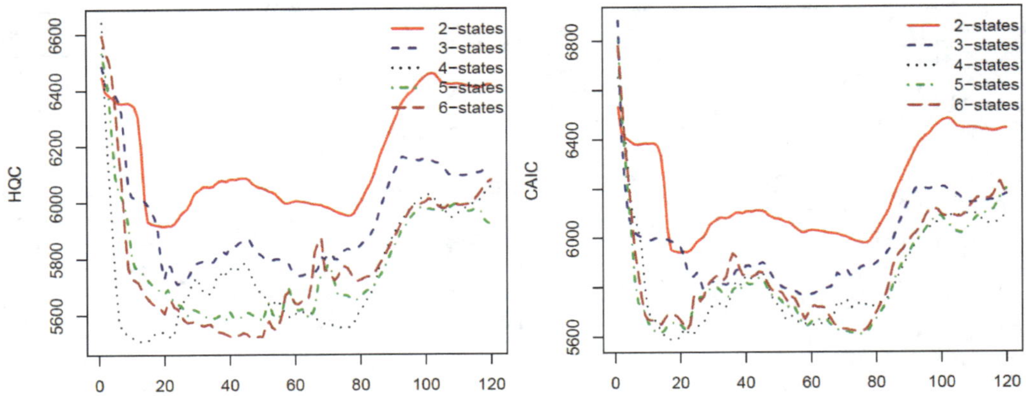

FIGURE 4.30
HQC (left) and CAIC (right) for 120 HMM parameter calibrations using S&P 500 monthly prices.

4.5.4 Out-of-Sample Stock Price Prediction

After choosing the best HMM for the S&P 500 index using these criteria above, the HMM with four states will be used to predict stock prices, compare the predictions with real stock prices, and then compare to the results of using the historical average return method. We will use the S&P 500 monthly prices from January 1950 to November 2016.

We first introduce how to predict stock prices using the HMM. The prediction process can be divided into three steps. First, the HMM parameters are calibrated using training data and by calculating the probability (or likelihood) of observing the data set. Then, we will find a "similar data set" in the past that has a similar likelihood to that of the training data. Finally, we will use the difference in stock prices in the last month of the founded sequence with the price of the consecutive month to predict future stock prices. This prediction approach is based on the work of [61]. However, the authors predicted daily prices for four airline stocks, while we predicted monthly prices for the U.S. benchmark market, the S&P 500. Suppose that we want to predict the closing price at time $T+1$ of the S&P 500. First, we will choose a fixed time length D of the training data (we call D the training window), then use the training data from time $T-D+1$ to T to calibrate the HMM parameters, λ. We assume that the observation probability $b_i(k)$, defined in Section 1.3.1, is Gaussian distribution, so the matrix B, in the parameter $\lambda = \{\mathbf{A}, \mathbf{B}, \mathbf{p}\}$, is a two by N matrix of means, μ, and variances, σ, of the N normal distributions, where N is the number of states. The initial HMM parameters for the calibration were calculated by using formula 4.24. The training data are the four sequences: "open", "low", "high" and "close" price:

$$O = \{O_t^{(1)}, O_t^{(2)}, O_t^{(3)}, O_t^{(4)}, t = T-D+1, T-D+2, ..., T\}.$$

We then calculate the probability of observation, $\mathbb{P}(O|\lambda)$. In the second step, we move the block of data backward by one month to have new observation data $O^{\text{new}} = \{O_t^{(1)}, O_t^{(2)}, O_t^{(3)}, O_t^{(4)}, t = T-D, T-D+1, ..., T-1\}$ and calculate $\mathbb{P}(O^{\text{new}}|\lambda)$. We keep moving blocks of data backward month by month until we find a data set O^*, $(O^* = \{O_t^{(1)}, O_t^{(2)}, O_t^{(3)}, O_t^{(4)}, t = T^*-D+1, T^*-D, ..., T^*\})$, such that $\mathbb{P}(O^*|\lambda) \simeq \mathbb{P}(\mathbf{O}|\lambda)$. In the final step, we estimate the stock's closing price for time $T+1$, $O_{T+1}^{(4)}$, by using the

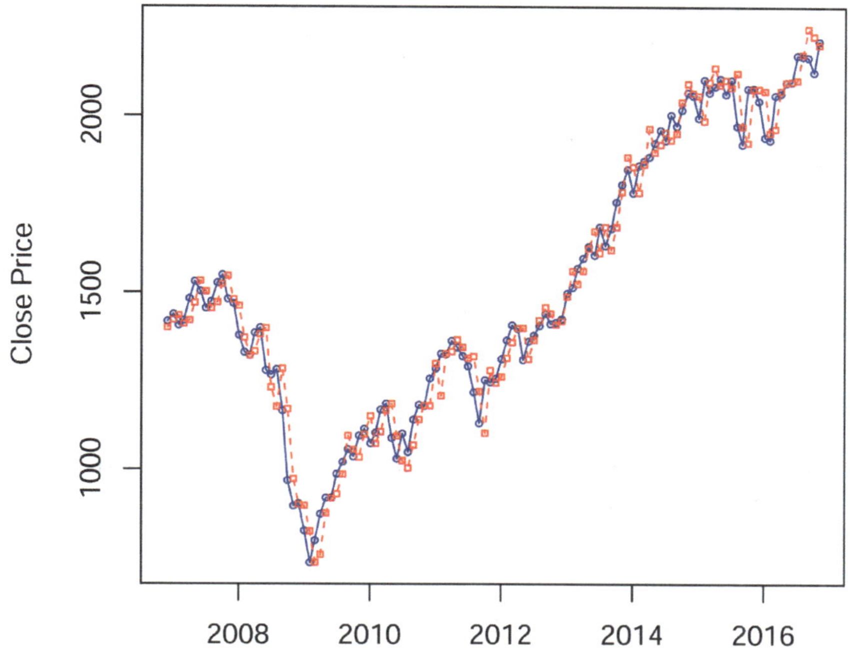

FIGURE 4.31
Predicted S&P 500 monthly prices from November, 2006 to November, 2016 using the four-state HMM.

following formula:

$$O_{T+1}^{(4)} = O_T^{(4)} + (O_{T^*+1}^{(4)} - O_{T^*}^{(4)}) \times \text{sign}(\mathbb{P}(O|\lambda) - \mathbb{P}(\mathbf{O}^*|\lambda)). \tag{4.25}$$

Similarly, to predict stock price at time $T+2$ we will use a new training data O: $O = \{O_t^{(1)}, O_t^{(2)}, O_t^{(3)}, O_t^{(4)}, t = T - D, T - D + 2, ..., T + 1\}$, to predict stock price for the time $T+2$. The calibrated HMM parameters, λ) in the first prediction were used as the initial parameters for the second prediction. We repeat the three-step-prediction process for the second prediction, and so on. For convenience, we use the training window D equal to the out-of-sample forecast period. In practice, we can choose an out-of-sample of any length. However, for the training window D we should determine a proper length base on the characteristics of chosen data to maximize the efficiency of model simulations.

Results of ten-year out-of-sample ($D = 120$) are presented in Figure 4.31, in which the S&P 500 historical data from January 1950 to October 2006 were used to predict its stock prices from November 2006 to November 2016. We can see from Figure 4.31 that the HMM effectively captures the price changes around the economic crisis time from 2008-2009.

Results of S&P 500 price predictions for other time periods are presented in Figures 4.32-4.33 bellow. It is hard to tell which training window length gives the most accurate prediction. We will calculate the predicted errors for each model to see which one is doing the best in the next section.

4.5.5 Model Validation

To test our predictions, we using the out-of-sample R^2 statistics, R^2_{OS}, introduced by [23]. Many researchers employed this statistic measure to evaluate their forecasting models. [23]

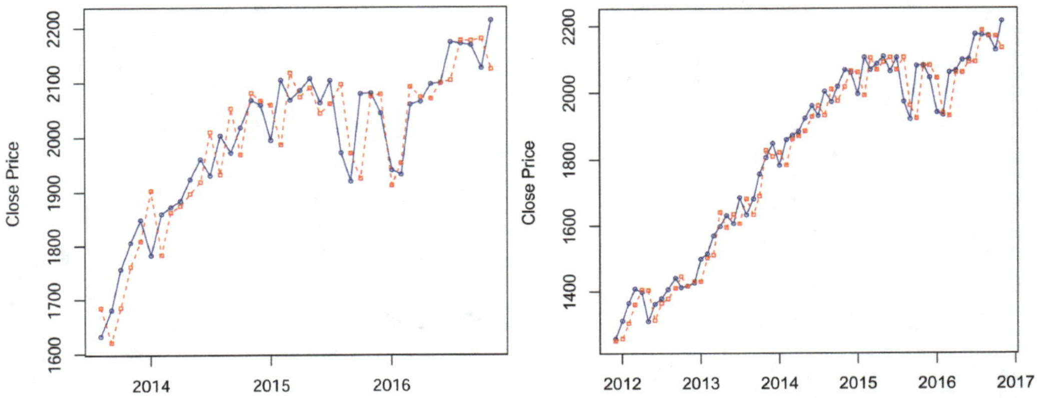

FIGURE 4.32
S&P 500 predicted monthly prices using the four-state HMM for 40 months or $D = 40$ (left) and 60-month or $D = 60$ (right).

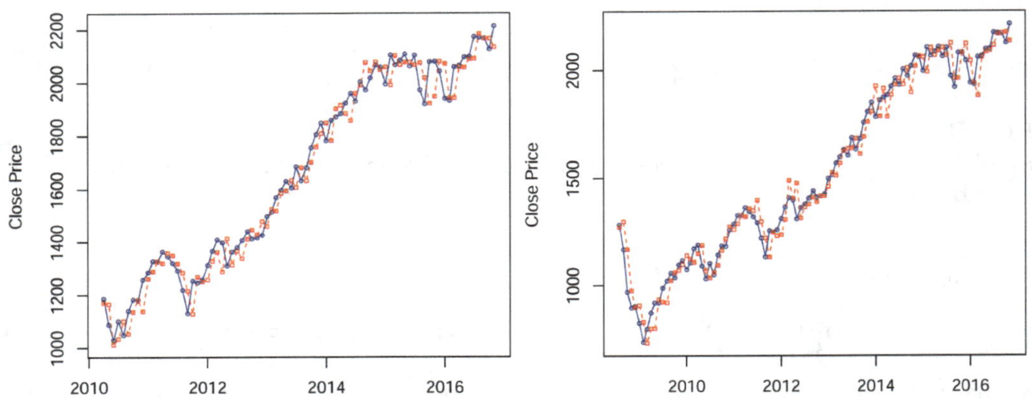

FIGURE 4.33
S&P 500 predicted prices using the four-state HMM for 80-months or $D = 80$ (left) and 100-months or $D = 100$ (right).

use the out-of-sample R^2_{OS} to compare the performances of the stock return prediction model inspired by [146] and that of the historical average return model. The authors used monthly total returns (including dividends) of the S&P 500. [118] use the out-of-sample R^2_{OS} to test the efficiency of their combination approach to the out-of-sample equity premium forecasting against the historical average return method. [163] also use the out-of-sample R^2_{OS} to show that their regime-switching combination forecasts of excess returns relative to the HAM. All of these above approaches were based on a regression model for multi-economic variables to predict stock returns.

The out-of-sample R^2_{OS} is calculated as follows. First, a time series of returns is divided into two sets: the first m points were called the in-of-sample data set, and the last q points were called the out-of-sample data set. Then, the out-of-sample R^r_{OS} for forecasted returns is defined as:

$$R^2_{\text{OS}} = 1 - \frac{\sum_{t=1}^{q}(r_{m+t} - \hat{r}_{m+t})^2}{\sum_{t=1}^{D}(r_{m+t} - \tilde{r}_{m+t})^2}, \tag{4.26}$$

where r_{m+t} is the real return at time $m+t$, $\hat{r}_{m+t}$ is the forecasted return from the desired model, and $\tilde{r}_{m+t}$ is the forecasted return from the competing model. We can see from the equation (4.26) that the R^2_{OS} evaluates the reduction in the mean squared predictive error (MSPE) of two models. Therefore, if the out-of-sample R^2_{OS} is positive then the desired model performed better than the competing model. The historical average return method is used as a benchmark competing model, which the forecasted return for the next time step is calculated as the average of historical returns up to the time,

$$\tilde{r}_{t+1} = \frac{1}{t}\sum_{i=1}^{t} r_i.$$

In this study, we use the HMM model to predict stock prices based on historical prices. However, we can calculate stock prices based on their predicted returns and reverse. Therefore, we will calculate the out-of-sample R^2_{OS} by using two approaches: out-of-sample R^2 for stock returns and out-of-sample R^2 for stock prices based on predicted returns (without dividends). Numerical results present in this section are for an out-of-sample approach and the training periods of the length $q = m = 60$.

The out-of-sample R^2 for relative return (without dividends), namely R^2_{OSR}, is determined by

$$R^2_{\text{OSR}} = 1 - \frac{\sum_{t=1}^{q}(r_{m+t} - \hat{r}_{m+t})^2}{\sum_{t=1}^{q}(r_{m+t} - \bar{r}_{m+t})^2}, \tag{4.27}$$

where r_{m+t} is the real relative stock return price at time $m+t$,

$$r_{m+t} = \frac{P_{m+t} - P_{m+t-1}}{P_{m+t-1}},$$

$\hat{r}_{m+t}$ is the forecasted relative return from the HMM,

$$\hat{r}_{m+t} = \frac{\hat{P}_{m+t} - P_{m+t-1}}{P_{m+t-1}} \tag{4.28}$$

and $\tilde{r}_{m+t}$ is the forecast price from the HAM, $\tilde{r}_{t+1} = \frac{1}{t}\sum_{i=1}^{t} r_i$. The out-of-sample R^2 for stock price based on predicted returns, namely R^2_{OSP}, is given by

$$R^2_{\text{OSP}} = 1 - \frac{\sum_{t=1}^{q}(P_{m+t} - \hat{P}_{m+t})^2}{\sum_{t=1}^{q}(P_{m+t} - \tilde{P}_{m+t})^2}, \tag{4.29}$$

where P_{m+t} is the real stock price at time $m+t$, $\hat{P}_{m+t}$ is the forecast price of the HMM, and $\tilde{P}_{m+t}$ is the forecast price based on the predicted return of the historical average return model,

$$\tilde{P}_{t+1} = P_t + P_t\tilde{r}_{t+1} = P_t(\tilde{r}_{t+1} + 1).$$

A positive R^2_{OSP} indicates that the HMM outperforms the HAM. Therefore, we will use the MSPE-adjusted hypothesized statistics test introduced by [31] to test the null hypothesis $R^2_{\text{OSP}} = 0$ vs. the alternate hypothesis $R^2_{\text{OSP}} > 0$. In the MSPE-adjusted test, we first define

$$f_{t+1} = (r_{t+1} - \tilde{r}_{t+1})^2 - [(r_{t+1} - \hat{r}_{t+1})^2 - (\tilde{r}_{t+1} - \hat{r}_{t+1})^2]. \tag{4.30}$$

We then test for $R^2_{\text{OSP}} = 0$ by regressing $\{f_{m+i}\}_{i=2}^{q-1}$ on a constant. Our p-value of testing coefficients of f_{m+i}, which are presented in Table 4.6, are bigger than the efficient level

TABLE 4.6
The mean squared predictive error adjusted test, MSPE-adj, for R^2_{OSP} and R^2_{OSR}

MSPE-adj.	Coefficients	Estimate	Std. Error	t-statistics	p-value	
R^2_{OSP}	Intercept	1.968e+02	3.358e-14	5.861e+15	2e-16	***
	f_{m+i}	2.181e-18	2.198e-17	9.900e-02	0.921	
R^2_{OSR}	Intercept	6.966e-05	3.561e-21	1.956e+16	<2e-16	***
	f_{m+i}	5.280e-19	6.807e-18	7.800e-02	0.938	

Note: '***' indicates that the test result is significant at the 0.001 level.

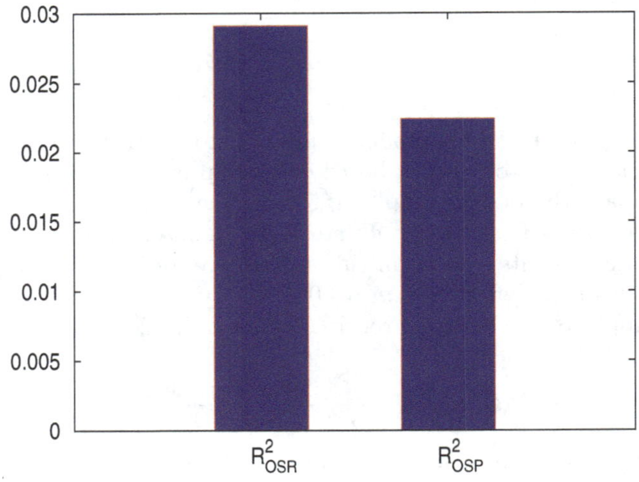

FIGURE 4.34
The R^2_{OSP}, and R^2_{OSR} of S&P 500 monthly prices from December 2011 to November 2016.

$\alpha = 0.001$, indicating that we support the null hypothesis that the coefficient of f_{m+i} for each test equals zero. Furthermore, the p-values of testing constant (intercept) for both tests are significant at $\alpha = 0.001$ level, which implies that we reject the null hypothesis that the $R^2_{\text{OSP}} = 0$ and support the alternate hypothesis that $R^2_{\text{OSP}} > 0$.

The out-of-sample R^2 for predicted prices and predicted returns are presented in Figure 4.34, showing that both out-of-sample R^2 are positive, that is, the HMM outperforms the historical average in predicting stock returns and stock prices.

Although the R^2_{OSP} compares the performances of two models on the whole out-of-sample forecasting period, it does not show the efficiency of the two competing models for each point estimation. Therefore, we use the cumulative squared predictive errors (CSPEs), which was presented by [163], to show the relative performances of the two models after each prediction. The CSPE statistic at time $m+t$, denoted by CSPE_t, is calculated as:

$$\text{CSPE}_t = \sum_{i=m+1}^{t} [(r_i - \tilde{r}_i)^2 - (r_i - \hat{r}_i)^2]. \tag{4.31}$$

From the definition of the cumulative squared predictive errors, we can see that if the function is increasing, the HMM outperforms the HAM. In contrast, if the function is decreasing, the HAM outweighs the HMM on the time interval. If we replace the return prices in 4.31 by the predicted prices, we will have the CSPE for prices. The CSPE of

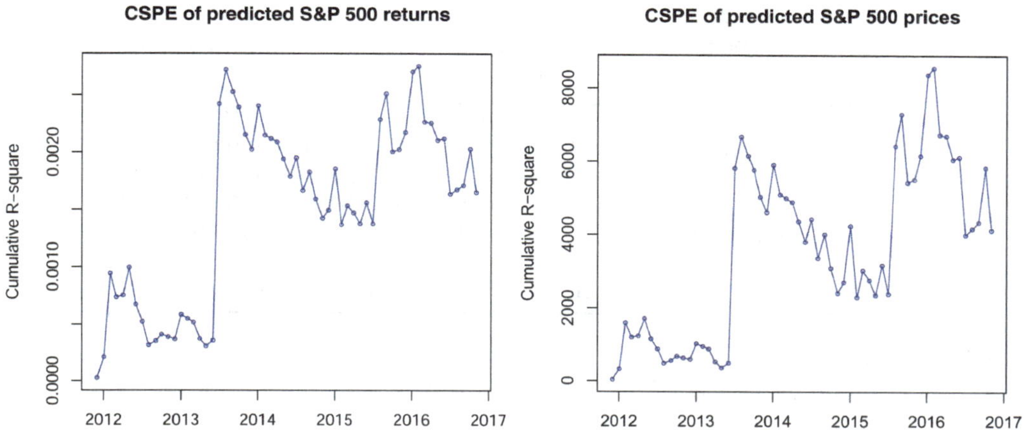

FIGURE 4.35
Cumulative squared predictive errors, CSPE, of S&P 500 monthly forecasted prices (right) and returns (left) from December 2011 to November 2016.

predicted returns and prices is presented in Figure 4.35. The results in Figure 4.35 show that although in some periods the HAR outperforms the HMM, the CSPEs are positive and follow an upward trend on the whole out-of-sample period. Therefore, we have a conclusion that in out-of-sample predictions, the HMM outperforms the HAR model.

We also compare the performances of the HMM with the HAM by using the four standard error estimators: the absolute percentage error (APE), the average absolute error (AAE), the average relative percentage error (ARPE), and the root-mean-square error (RMSE). These error estimators are calculated using the following expressions

$$\text{APE} = \frac{1}{\bar{r}} \sum_{i=1}^{N} \frac{|r_i - \tilde{r}_i|}{N} \tag{4.32}$$

$$\text{AAE} = \sum_{i=1}^{N} \frac{|r_i - \tilde{r}_i|}{N} \tag{4.33}$$

$$\text{ARPE} = \frac{1}{N} \sum_{i=1}^{N} \frac{|r_i - \tilde{r}_i|}{N} \tag{4.34}$$

$$\text{RMSE} = \sqrt{\frac{1}{N} \sum_{i=1}^{N} (r_i - \tilde{r}_i)^2}, \tag{4.35}$$

where N is a number of simulated points, r_i is the real stock price (or stock return), $\tilde{r}_i$ is the estimated price (or return), and $\bar{r}$ is the mean of the sample. We use the error estimators to calculate the errors of predictions of the HMM and HAR models for predicted returns in equation (4.27) and predicted prices in equation (4.29), respectively.

Adopting the R^2_{OS} statistics, we define the efficiency measure to compare the HMM with the HAR model as:

$$\text{Eff} = 1 - \frac{\text{HMM Error}}{\text{HAR Error}}. \tag{4.36}$$

Errors of the two models and the efficiency of the HMM over the HAR are calculated using equations (4.49)-(4.36). A positive efficiency (Eff) indicates that the HMM surpasses the

TABLE 4.7
Error and efficiency of S&P 500's predicted prices and predicted returns using the HMM vs. HAR model

Error Estimators	Models	APE	AAE	ARPE	RMSE
Predicted Return	HMM	0.3325	0.0245	0.0303	2.4371
	HAR	0.1708	0.0249	0.0308	2.4752
	Eff.	**-0.9467**	**0.0161**	**0.0161**	**0.0154**
Predicted Price	HMM	0.0242	43.5080	54.8606	0.0240
	HAR	0.0246	44.2091	55.4853	0.02440
	Eff.	**0.0163**	**0.0159**	**0.0113**	**0.0164**

TABLE 4.8
S&P 500 trading using the HMM, HAR, and Buy & Hold models for different time periods

Trading Period	Model	Investment ($)	Earning ($)	Cost ($)	Profit (%)
40 months	HMM	168,155	65,172	168	38.66
(7/2013	HAR	168,573	52,762	14	31.29
-11/2016)	Buy & Hold	168,573	52,762	14	31.29
60 months	HMM	124,696	95,205	378	76.05
(12/2011	HAR	131,241	90,094	14	68.64
-11/2016)	Buy & Hold	124,696	96,639	14	77.49
80 months	HMM	116,943	113,971	392	97.12
(4/2010	HAR	116,943	104,392	14	89.26
-11/2016)	Buy & Hold	116,943	104,392	14	89.26
100 months	HMM	126,738	94282	616	73.91
(8/2008	MAR	126,738	73,010	84	57.54
-11/2016)	Buy & Hold	126,738	94,597	14	74.63
120 months	HMM	141,830	100,614	770	70.39
(11/2006	HAR	140,063	81,272	14	58.15
-11/2016)	Buy & Hold	140,063	81,272	14	58.15

HAR model. Table 4.7 presents errors and efficiency of the HMM over the HAR model. We can see from the table that the HAR model beats the HMM based on the absolute percentage error estimator, APE, for stock returns. However, in all the remaining cases, we have positive efficiency, which strongly indicates that the HMM outperforms the HAR.

4.5.6 Stock Trading Using an HMM

In this section, we will use the predicted returns of the HMM and the HAR models to trade S&P 500 using monthly data. If the predicted stock return is positive for the next month, we will buy the stock this month and then sell it if the next predicted return is negative. We assume that we buy and sell with closing prices. If the HMM predicts that the stock price will not increase next month then we will do nothing. We also assume that trading cost is $7.00 per trade, this assumption is based on the trading fee on the U.S. market. The trading fees vary from broker to broker; some brokers even offer free trading for qualified investors. In reality, we will have many different ways to minimize the transaction fee. One simple way is increasing the volume or number of shares in each trade. For each trade, we will buy or sell 100 shares of each of the S&P 500. Based on the results of model selections in Section 4.5.3, we only use the HMM with four states for the stock trading. We use different training and trading periods to test our model. Trading results of using the HMM and the HAR models are presented in Table 4.8.

In Table 4.8, the "Investment" is the price that we bought 100 shares of the S&P 500 index at the first time. The "Cost" was calculated by multiplying the total numbers of "buy" and "sell" during the period by $7.00. The "Profit" is the percentage of return after trading cost. In this table, we list the trading results of the HMM, the HAR models, and the Buy & Hold method. In the Buy & Hold method, we assume that an investor will buy 100 shares of the S&P 500 at the beginning of the trading period and hold it until the end of the period. The results show that the HMM beats the HAR model for all cases and outperforms the Buy & Hold technique for most of the cases except for 100 month period. In three cases: 40 months, 80 months, and 120 months, the HAR and Buy & Hold models have identical results since all of the HAR's predicted returns are positive. One disadvantage of the HAR model is due to its predictions depending solely on the mean of historical data, which is not sensitive to a price change at a point. In contrast, the HMM strongly captures the change of input data at a single point. Therefore, the HMM is more suitable for stock price forecasting than the HAM.

4.5.7 Conclusions

Stock performances are an essential indicator of the strengths and weaknesses of the stock's corporation and the economy in general. In this book, we have used the HMM to predict monthly closing prices of the S&P 500 and then used these predictions to trade the stock. We first use the four criteria: AIC, BIC, HQC, and CAIC to examine the performances of the HMM with numbers of states from two to six. The results show that the HMM with four states is the best model among these five HMMs. We then use the four-state HMM to predict monthly prices of the S&P 500 and compare with the results obtained by the benchmark historical average return model. We use the out-of-sample R^2, and other model validation methods to test our model. The results show that the HMM outweighs the historical average method. After validating the HMM model for out-of-sample predictions, we apply the model to trade the S&P 500 using different training and testing periods. The numerical results show that the four-state HMM outperforms the HAR not only in return predictions but also in stock trading. Overall, the HMM model beats the Buy & Hold strategy and yields higher percentage returns. The testing and trading results indicate that the HMM is the potential model for stock price predictions and stock trading.

4.6 Enhancing the Hull-White Short Rate Model with an HMM

Our work presents a modeling study of short term interest rates using the HMM and the Hull-White (HW) model. For this purpose, we modify the original HW model by adding a regime variable to its instantaneous forward function. This variable is defined by the regimes of the short term interest rate which are found by a two-state HMM. This combination of the HMM and the HW model for generating interest rate predictions results in a significant improvement, reducing the error of the estimations by about 50% compared to that of using the HW model alone. Furthermore, the errors in the simulations using the HMM and the HW models have a smaller standard deviation when compared to just using the HW model. Adjusted R^2 results also show that the regime variable is significant. This improvement to the short-term interest rate model has a substantial impact on financial economics and related fields.

4.6.1 Introduction

The short-term interest rate (short-rate) is an important input for many financial models, especially for financial derivative models. However, the short-rate is difficult to model accurately. The famous Black-Scholes Model for pricing options is based upon an assumption of a constant short-rate. However, in reality, the short rate is not constant; it fluctuates as economic conditions change. Classical short rate models, as in [33, 138] assume the short rate follows a Brownian motion process, which is not supported by empirical evidence. Some of limitations of these early models were addressed in [66] with the development of the Hull and White Extended Vasicek and Hull and White Extended CIR models. These extensions add more flexibility to the short interest rate models. However, we also experience jumps in the time series of US real interest rates as a result of important macroeconomic changes, the most recent change being the 2008 economic crisis. Researchers tried to accommodate these jumps by allowing for the possibility of regime-switching in short-term interest rate models. Garcia and Perron [53] modeled the short rate in this manner. These approaches added too much structure to short rate models which most of the time resulted in data overfitting. Also, the estimations of these model were excessively expensive. The forward rate is the most important component of the Hull-White model, but has largely been ignored in research. Typically, the forward curve is assumed to be a function of time. However, it is also subject to regime changes due to changes in the macroeconomic environment. Adding dummy regime variables to the forward function is a simple relaxation of the Hull-While model. In this book, we consider the influence of regimes of the forward rate on the short-term interest rate in the Hull-White model. First, we use the HMM to find regimes of the interest rate and code these regimes as dummy variables in the prediction of the forward rate used to generate the short-term interest rate. Using the HMM to find regime variables to add to the forward rate function of the HW model is a simple modification, but has a significant impact. This modification will allow for more accurate short rate modeling, which is critical in the pricing of annuity contracts, investment outcomes and decisions, and in other relevant models in actuarial science and financial mathematics.

The remainder of this study is organized as follows. Section 4.6.2 presents an overview of the HMM. Section 4.6.3 provides a brief introduction of the HW model. The methodology and results of using the HMM and HW model to simulate the short term interest are described in Section 4.6.4. Section 4.6.5 concludes the study.

4.6.2 Overview of the HMM

The HMM was developed by Baum and Petrie in 1966 [12]. This is a stochastic signal model based upon the following assumptions:

- The observation at time t, O_t, was generated by some process whose state, S_t, is hidden.
- The hidden states satisfy the first-order Markov property.
- The transition matrix $\mathbf{A} = (\mathbf{a_{ij}})$ is constant, where a_{ij} is the probability of being in state S_j at time t given that the observation at time $t-1$ is in state S_i.
- The observation at time t of an HMM has a particular probability distribution corresponding to a possible state. The observation probability matrix is denoted by $\mathbf{B}$.

The parameters of the HMM are the matrices $\mathbf{A}$ and $\mathbf{B}$ and the vector $\mathbf{p}$. For convenience, we use compact notation for the parameters:

$$\lambda \equiv \{\mathbf{A}, \mathbf{B}, \mathbf{p}\}.$$

If the observation probability assumes the Gaussian distribution, then $b_i(k) = b_i(O_t = v_k) = \mathcal{N}(v_k, \mu_i, \sigma_i)$, where μ_i and σ_i are the mean and variance of the distribution corresponding to the state S_i, respectively, and $\mathcal{N}$ is Gaussian density function. In this case, the parameters of the HMM are

$$\lambda \equiv \{\mathbf{A}, \mu, \sigma, \mathbf{p}\}.$$

Since its introduction in 1966, the HMM underwent many developments. The innovations primarily involved calibration of the model's parameters. Baum and his colleagues published a maximization method to calibrate the HMM parameters using a single observation sequence in [13]. Sondhi introduced a maximum likelihood estimation method for the HMM with multiple observation sequences, assuming that all the observation sequences are independent in [86]. In 2000, Li, Parizeau, and Plamondon [88] presented an HMM training for multiple observation sequences without the assumption of independence of these sequences. More details about this model and its algorithms can be found in [115] and [105].

The HMM finds applications in many different areas such as speech recognition systems, computational molecular biology, and financial market predictions. Researchers have applied the HMM to forecast stock prices. Hassan and Nath [61] used the HMM to predict the stock price for interrelated markets. Kritzman, Page, and Turkington [79] applied the HMM with two states to predict regimes in market turbulence, inflation, and industrial production index. Guidolin and Timmermann [58] used the HMM with four states and multiple observations to study asset allocation decisions based upon regime switching in asset returns. Ang and Bekaert [5] applied the RSM (another name for the HMM) for international asset allocation. Nguyen [103] used the HMM with both single and multiple observations to forecast economic regimes and stock prices. B. Nobakht, C. E. Joseph and B. Loni [108] implemented the HMM by using multiple observation sequences (open, close, low, high) prices of a stock to predict its closing price. Nguyen and Nguyen [105] used the HMM for single observation data to predict regimes of some economic indicators and made stock selections based upon the performances of these stocks during the predicted regimes.

Recently, Elliott and Wilson [42] used the HMM to model the short-term interest rate by assuming that the mean-reverting level follows a finite-state, continuous-time Markov chain. Erlwein and Mamon [44] implemented the HMM for the interest rate model presented in [42] for a financial data set of 30-day Canadian Treasury bill yields. In this section, we combine the HMM with a standard model for the interest rate, the Hull-White model, to generate predictions for the short-term interest rate and compare the results with the results obtained by only using the Hull-White model.

4.6.3 Overview of Hull-White Model

The short-term interest rate plays a key role in the modeling of financial securities, derivatives, and other interest rate contingent claims. Beginning with Vasicek [138], many models for the short-rate have been proposed. Each model has its advantages and limitations. Two of the earliest models were developed in [33] and [138]. Some of the limitations of these early models were addressed by Hull and White [66] with the development of the Hull and White Extended Vasicek and Hull and White Extended CIR models. We provide here a brief description of the Vasicek, CIR, and Hull and White models, while a more detailed description of these and other models for the short-rate can be found in [20].

In the Vasicek model, the risk neutral dynamics are given by the stochastic differential equation

$$dr(t) = k(\theta - r(t))\ dt + \sigma dW(t)$$

with $r(0) = r_0$ and $k, \theta, \sigma, r(0)$ all positive constants. This model is mean-reverting to θ and implies a normally distributed short-rate. Because the short-rate is normally distributed, the Vasicek model is analytically tractable and produces analytical formulas for bond and option prices. The major drawbacks of the Vasicek model are its inability to match the term structure observed in the market and the possibility of predicting negative rates.

To address and eliminate the possibility of negative rates, Cox, Ingersoll, and Ross [33] proposed a short-rate model of the form

$$dr(t) = k(\theta - r(t))\, dt + \sigma\sqrt{r(t)}\, dW(t)$$

with $r(0) = r_0 > 0$ and where k, θ, σ, and $r(0)$ are all positive constants. This CIR model shares the advantage of analytic tractability as the short-rate in this model has a non-central chi-square distribution. The CIR model also has the advantage of ensuring predicted rates will never be negative. However, it shares the drawback of the Vasicek model that the term structure of the short-term rate can not be matched.

To remedy this issue with the Vasicek and CIR models, Hull and White [66] proposed extensions of both models that are consistent with the current term structure of interest rates. We will examine a version of the Hull-White Extended Vasicek model, as it is the focus of this study and has the advantage of being analytically tractable. The stochastic differential equation describing the Hull-White Extended Vasicek model is

$$dr(t) = (\theta(t) - a(t)r(t))\, dt + \sigma(t)dW(t)$$

where $a(t)$, $\sigma(t)$, and $\theta(t)$ are deterministic functions of time, $\theta(t)$ is chosen to fit the initial term structure of the interest rates, and $\sigma(t)$ is chosen to fit to the current spot or forward-rate volatility term structure. As noted in [66] and [20], if an exact calibration to the term structure of interest rates is desired, then the perfect fitting of the volatility structure may be problematic. Because of this, we follow the development of a simplified model, analyzed by Hull and White [66], which has the form

$$dr(t) = (\theta(t) - ar(t))\, dt + \sigma dW(t) \tag{4.37}$$

where a and σ are positive constants and $\theta(t)$ is chosen to precisely fit the current term structure of interest rates observed in the market. We will refer to this as the Hull-White one-factor model. Because this model is an extension of the Vasicek model, it too implies a normal distribution for the short rate process $r(t)$ and thus is analytically tractable. However, it also has the drawback of the possibility of producing negative rates.

Let $(\Omega, \mathcal{F}, Q)$ be a probability space. Define the market instantaneous forward rate at time 0 for maturity T as

$$f^M(0,T) = -\frac{\partial \ln P^M(0,T)}{\partial T},$$

where $P(t,T)$ is defined in 4.38.

Let $E(\cdot|\mathcal{F}_s)$ and $\text{Var}(\cdot|\mathcal{F}_s)$ be the expectation and variance conditional on the $\mathcal{F}_s$ σ-field, respectively. Then it can be shown that the Hull-White Model implies a normally distributed short-rate process with

$$E(r(t)|\mathcal{F}_s) = r(s)e^{-a(t-s)} + \alpha(t) - \alpha(s)e^{-a(t-s)}$$

where

$$\alpha(t) = f^M(0,t) + \frac{\sigma^2}{2a^2}[1 - e^{-2a(t-s)}]^2$$

and

$$\text{Var}(r(t)|\mathcal{F}_s) = \frac{\sigma^2}{2a}[1 - e^{-2a(t-s)}].$$

The Hull-White Model also gives closed formulas for the price of a zero-coupon bond issued at time t maturing at time T, $P(t,T)$. The price is given by

$$P(t,T) = A(t,T)e^{-B(t,T)r(t)} \tag{4.38}$$

with

$$A(t,T) = \frac{P^M(0,T)}{P^M(0,t)} \exp\left[B(t,T)f^M(0,t) - \frac{\sigma^2}{4a}(1-e^{-2at})B(t,T)^2\right]$$

and

$$B(t,T) = \frac{1}{a}[1-e^{-a(T-t)}].$$

4.6.4 Combining the HMM with the Hull-White One-Factor Model

In this section we explain how to use the Hull-White one factor model and the HMM to simulate the interest rate. We first present the methodology, followed by the data preparation and numerical results.

4.6.4.1 Methodology

Consider the time interval $t_0, t_1, \ldots, t_N$ with equal width $\Delta_t = \frac{1}{N}$. The short interest rate series $r_{t_0}, r_{t_1}, \ldots, r_{t_N}$ can be simulated using the Hull-White one factor model (4.37) recursively with an initial interest rate r_0:

$$r_{t_i} = \theta(t_{i-1})\Delta t + (1-a\Delta t)r_{t_{i-1}} + \sigma\sqrt{\Delta t}Z_{t_i}, \tag{4.39}$$

where the deterministic function $\theta(t)$ is chosen as

$$\theta(t) = \frac{\partial}{\partial t}f(0,t) + af(0,t) + \frac{\sigma^2}{2a}(1-e^{-2at}). \tag{4.40}$$

Hull and White [66] showed that we can eliminate the term $\frac{\sigma^2}{2a}(1-e^{-2at})$ in equation (4.40) since its value is fairly small. Therefore, we just use

$$\theta(t) = \frac{\partial}{\partial t}f(0,t) + af(0,t). \tag{4.41}$$

In the HW, the market instantaneous forward function $f(0,t)$, the forward rate at time 0 for maturity t, was assumed to be a third degree polynomial function of time t of the form

$$f(0,t) = b_0 + b_1.t + b_2.t^2 + b_3.t^3. \tag{4.42}$$

In the HMM-HW model, we add a state variable for the interest rate, denoted λ, into the instantaneous forward function defined as a dummy variable. Thus, the forward function has the form

$$f(0,t) = c.\lambda + b_0 + b_1.t + b_2.t^2 + b_3.t^3. \tag{4.43}$$

First, we use the HMM to find the states of the interest rate, the dummy variable λ of the forward rate function, then use polynomial regression to find coefficients of forward function in equation (4.43). Using equations (4.41) and (4.44) gives

$$r_{t_i} - r_{t_{i-1}} = \frac{\partial}{\partial t}f(0,t_{i-1})\Delta t + a(f(0,t_{i-1}) - r_{t_{i-1}})\Delta t + \sigma\sqrt{\Delta t}Z_{t_i}. \tag{4.44}$$

Thus

$$\sigma\sqrt{\Delta t}Z_{t_i} = r_{t_i} - r_{t_{i-1}} - \frac{\partial}{\partial t}f(0,t_{i-1})\Delta t + a(f(0,t_{i-1}) - r_{t_{i-1}})\Delta t.$$

We then find a as a minimizer of function

$$HW(a) = \sum_{i=1}^{n}(r_{t_i} - r_{t_{i-1}} - \frac{\partial}{\partial t}f(0, t_{i-1})\Delta t + a(f(0, t_{i-1}) - r_{t_{i-1}})\Delta t)^2. \tag{4.45}$$

The last parameter that we need to calibrate is σ. To find σ, we first predict the interest rate $\hat{r}$ using the formula

$$\hat{r}_i = \left(\frac{\partial}{\partial t}f(0, t_{i-1}) + af(0, t_i)\right)\Delta t + (1 - a\Delta t)r_{t_{i-1}}, \tag{4.46}$$

for $i = 1, 2, \ldots, N$, and $\hat{r}_{t_0} = r_{t_0}$. Then

$$\sigma = \frac{s}{\sqrt{\Delta t}}, \tag{4.47}$$

where s is standard deviation of error series $\hat{r}_{t_i} - r_{t_i}, i = 0, 1, \ldots, N$.

After calibrating all the needed parameters, we then simulate the interest rate using Equation (4.44). Data and results of the simulations will be presented in the next section.

4.6.4.2 Implementation and Results

We use daily data on the one-year interest rate in this study. The economic crisis in 2008 had huge effects on many economic indicators, including the interest rate. Therefore, we want to use the HMM to detect the crisis time and find regimes of the interest rate. We use the HMM with two regimes, which represent the two main economic conditions: growth and recession. We assume that the observed probability distributions corresponding to these two regimes are Gaussian in nature. Daily data of the one-year interest rate from December 31, 2004 to January 1, 2010 is used to calibrate parameters and find the data's regimes using Baum-Welch and Viterbi algorithms. First, we use the Baum-Welch algorithm to calibrate the model parameters $\lambda = \{\mathbf{A}, \mu, \sigma, \mathbf{p}\}$. Then, we calculate the ratio of the means and variances of the two normal distributions, $\frac{\mu_i}{\sigma_i}$, $i = 1, 2$, and define regime one as the regime of the normal distribution with a higher ratio. The trained transition probability matrix is

$$\mathbf{A} = \begin{bmatrix} 1.0000 & 0.0000 \\ 0.0013 & 0.9987 \end{bmatrix}.$$

The initial probability of being in regime one or two at time 0 is $\mathbf{p} = (0, 1)$. The parameters of the two normal distributions corresponding to the two regimes are: ($\mu_1 = 4.34$, $\sigma_1 = 0.73$) and ($\mu_2 = 0.84$, $\sigma_2 = 0.72$). The regimes of the short term interest rate from December 31, 2004 to January 1, 2010 are presented in Figure 4.36.

The result shows that the interest rate in the period had two regime periods: regime one from December 31, 2004 to December 31, 2007 and regime two from January 1, 2008 to January 1, 2010. We pick a period of ten months before the regime change on December 31, 2007 and ten months after the regime change to make a pool of data for the interest rate simulations. We choose to expand the data to ten months before and after the regime change date because we want a pool of data that has only two time periods consisting of the first regime and the second regime. The transition date between these two regimes is December 31, 2007. The number of points in the data pool is the number of business days during the period, which is 450 days. Each year has 252 working days. Therefore, we choose randomly one time series of 252 consecutive points from the pool of data to train parameters for the Hull-White model and use the trained parameters to simulate the interest rate. Thus, the time step in the HW model is $\delta_t = 1/252$. Since we used the HMM to find regimes of

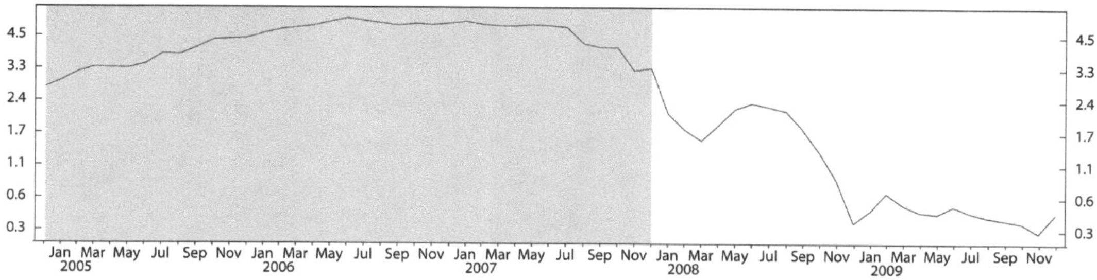

FIGURE 4.36
Predicted regimes of one-year interest rate using an HMM.

the data pool, for any random one-year-period series in the data pool we already know its regimes which were switched on December 31, 2007. We follow the methodology presented in the previous section to simulate the one-year interest rate using a random daily time series from the pool starting on October 30, 2007 to October 28, 2008. The regime switch date corresponds to the time variable $t = 42/252$. First, based on its regimes found by using the HMM, we determine the values of a regime (or dummy) variable λ_t in equation (4.43)

$$\lambda_t = \begin{cases} 1 & \text{if } 0 \leq t \leq 42/252 \\ 0 & \text{if } 42/252 < t \leq 1. \end{cases} \tag{4.48}$$

We then use polynomial regression to determine the instantaneous forward rate function represented by Equation (4.43):

$$f(0,t) = -0.8925\lambda_t + 4.0214 - 9.1573t + 20.6133t^2 - 13.3839t^3.$$

Thus, the derivative of $f(0,t)$ with respect to t is:

$$f'(0,t) = -9.1573 + 41.2266t - 40.1517t^2.$$

We substitute these functions into the function $HW(a)$ in equation (4.45) and use least squares regression to optimize the function and find its minimizer a. Then, we predict the sequence $\hat{r}$ using equation (4.46) and calibrate the parameter σ using equation (4.47). Finally, we compute the function $\theta(t)$ using Equation (4.41). We have the parameters $a = 27.5136$, $\sigma = 1.4797$, and

$$\begin{aligned}\theta(t) = &- 9.1573 + 41.2266t - 40.1517t^2 + 27.5136\times \\ &(-0.8925\lambda_t + 4.0214 - 9.1573t + 20.6133t^2 - 13.3839t^3).\end{aligned}$$

After determining all of HW's parameters $(a, \sigma, \theta(t))$, we use the parameters to simulate the interest rate for the same period using Equation (4.44). The results of the simulation are presented in Figure 4.37. In Figure 4.37, we plot the three series: real interest rate, the simulated interest rate using the original HW model, and the simulated interest rate using the HMM-HW model. We compute errors of simulations using the absolute percentage error (APE), the average absolute error (AAE), the average relative percentage error (ARPE), and the root-mean-square error (RMSE). These error estimators are calculated using the formulas

$$\text{APE} = \frac{1}{\bar{r}} \sum_{i=1}^{N} \frac{|r_i - \tilde{r}_i|}{N}, \tag{4.49}$$

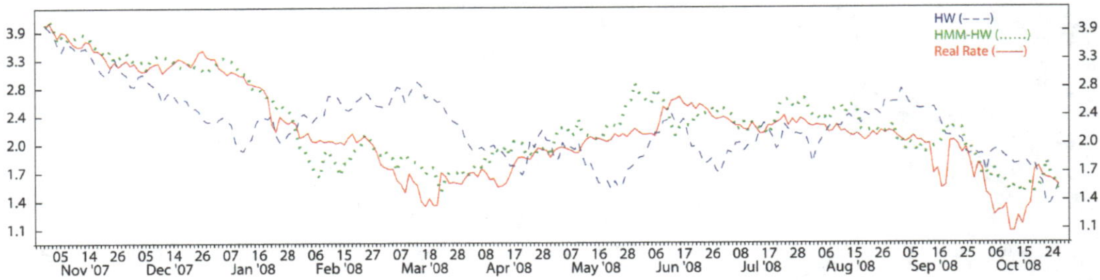

FIGURE 4.37
Simulated interest rate using the HW and HMM-HW models.

TABLE 4.9
Errors of simulating interest rate using the HW and HMM-HW models

Error Estimator	AAE	RMSE	ARPE	APE
HW	0.4617	0.5899	0.0018	0.2034
HMM-HW	0.2150	0.2778	0.0009	0.0947

$$\text{AAE} = \sum_{i=1}^{N} \frac{|r_i - \tilde{r}_i|}{N}, \tag{4.50}$$

$$\text{ARPE} = \frac{1}{N} \sum_{i=1}^{N} \frac{|r_i - \tilde{r}_i|}{N}, \tag{4.51}$$

and

$$\text{RMSE} = \sqrt{\frac{1}{N} \sum_{i=1}^{N} (r_i - \tilde{r}_i)^2}, \tag{4.52}$$

where N is a number of simulated points, r_i is the real interest rate, $\tilde{r}_i$ is the simulated rate, and $\overline{r}$ is the mean of the sample. The errors are listed in Table 4.9. The results in Table 4.9 show that the HMM-HW reduces errors of the interest simulations by about 50%. We repeat the process for 1500 random series in the pool and plot the histogram of the errors, calculated by Eq. (4.49), (4.50), (4.51), and (4.52), in Figures 4.38-4.41, respectively.

The results in Figures 4.38-4.41 show that the HMM-HW model reduces not only the mean of the errors but also their variance. We see from the figures that the histogram of the HMM-HW errors is shifted to the left compared with the histogram of the HW errors.

Furthermore, the distributions of errors of these two methods for the interest rate have positive skewness and their kurtosis is far from the kurtosis of the normal distribution. We also test the efficiency of adding the regime (dummy) variable to the HW by calculating the adjusted R^2 of regressions and plot the results in Figure 4.42. In Figure 4.42, the horizontal axis is the end date of each series of these 1500 simulations and the vertical axis is the adjusted R^2. The adjusted R^2 was calculated by using the formula

$$R^2_{adj} = 1 - \frac{(1 - R^2)(n - 1)}{n - k - 1},$$

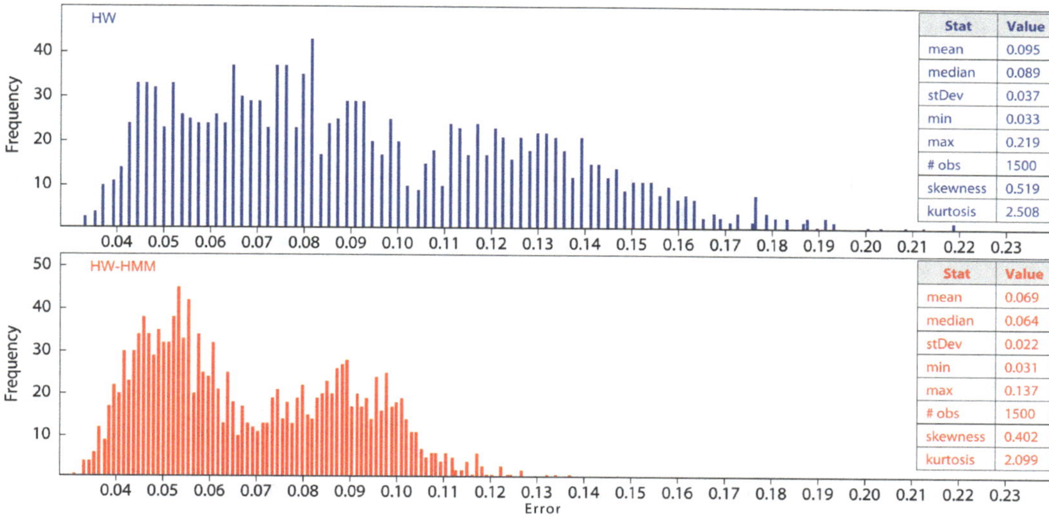

FIGURE 4.38
APE errors of interest rate simulations using the HW and HMM-HW models.

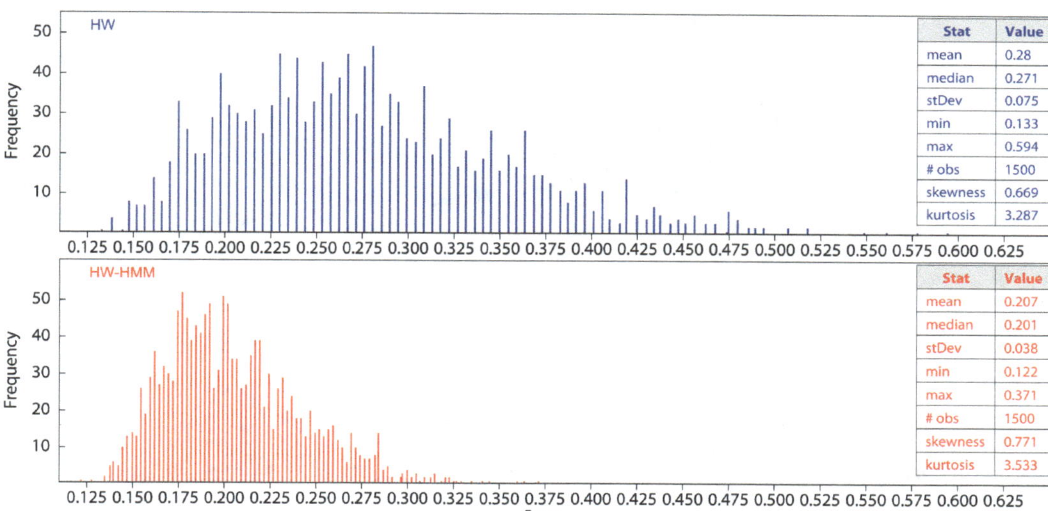

FIGURE 4.39
AAE errors of interest rate simulations using the HW and HMM-HW models.

where N is sample size and k is the number of variables in the model, excluding the constant. If a useless variable is added to a regression model, the adjusted R^2 will decrease. In contrast, if we add a useful variable to the model, adjusted R^2 will increase. The results show that the regime variable is a significant variable in the forward function. The adjusted R^2 increased when we added the dummy variable to the model.

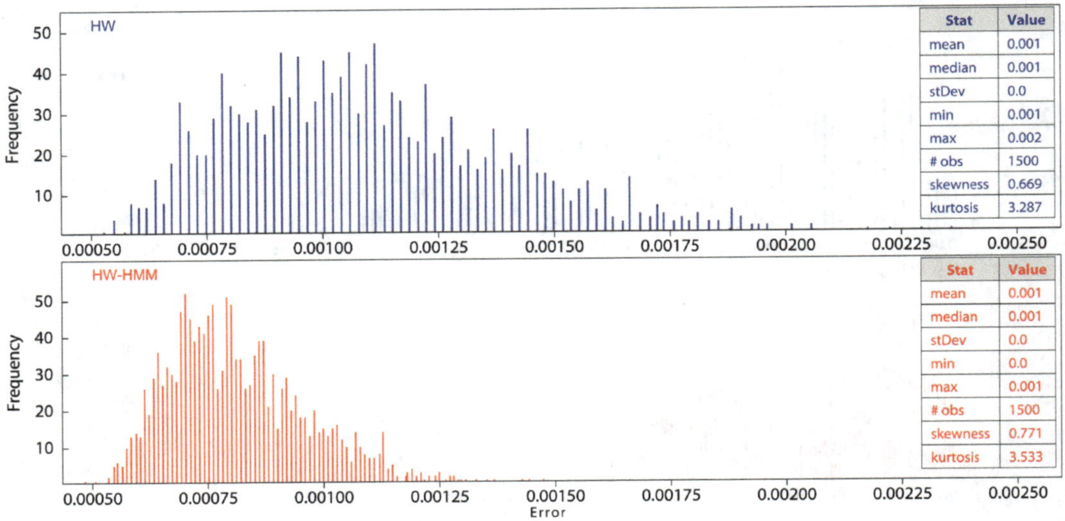

FIGURE 4.40
ARPE errors of interest rate simulations using the HW and HMM-HW models.

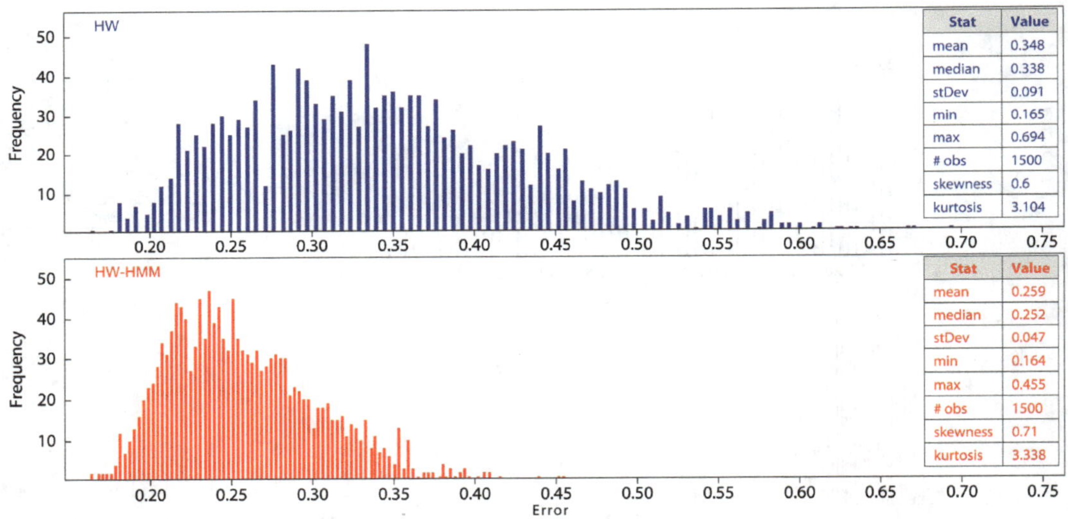

FIGURE 4.41
RMSE errors of interest rate simulations using the HW and HMM-HW models.

4.6.5 Conclusions

The Hull-White one-factor model is a classical model for the short-term interest rate. The forward rate function in the HW model is assumed to be a function of time t. However, we found jumps in the time series of U.S. real interest rates as a result of important macroeconomic changes, most recently during the 2008 economic crisis. Therefore, we investigated the effects of the jumps on short term interest rate modeling.

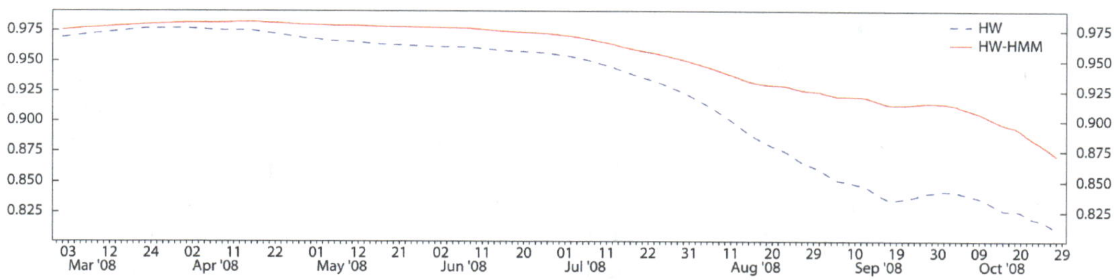

FIGURE 4.42
Adjusted R^2 of regression of forward function of the HW and HMM-HW models.

In this study, we added a regime variable to the forward rate function to make it more adapt to the regime switch of the interest rate. The regimes were identified by the two-state (or two-regime) HMM. Adding the dummy regime variable to the forward function is a simple relaxation of the Hull-White model. The modification made a significant improvement to the HW model. The error of the interest rate estimations using the HMM and the HW (HMM-HW) model was reduced by half compared with the original HW model. Furthermore, the results of the short term interest rate simulations showed that the HMM-HW histogram errors were shifted to the left and shrunk horizontally by half of the size of the HW's histogram errors. Hence, using the HMM-HW model for simulating the short term interest rates reduced not only the mean but also the variance of errors. Additionally, the adjusted R^2 of the regression model for the forward rate function was increased when we added the regime variable. All of these results indicate that the regime variable has a significant influence in modeling the interest rate. This important improvement of the short-term interest rate model has a substantial impact on the pricing of annuity contracts, investment outcomes and decisions, and in other relevant models in actuarial science and financial mathematics.

4.7 HMM for Global Stock Portfolio Management

The HMM is a powerful unsupervised machine learning model that has been used to detect the regimes of various observation data, especially for time series. In this section, we use this model to select stocks from the global stock market. This procedure involves some well-defined steps. First, five important factors of a stock are identified, and a score is assigned for each of the factors based on its historical performances. Second, the HMM is used to predict the regimes of six global economics indicators and find the time periods in the past during which these indicators have a combination of regimes that is similar to those predicted. Then, we analyze the five stock factors of the All Country World Index in the identified time periods to assign a weighted score for each stock factor and to calculate the composite score of the five factors. Finally, we make a monthly selection of 10% of the global stocks which have the highest composite scores. We compare the performances of our portfolio with those of the benchmark global index and the equal weight model.

4.7.1 Introduction

The HMM has been used in in many different fields such as speech recognition, bio-informatics, weather forecast, and finance. Recently, researchers have applied the HMM in financial economics. The HMM was used to forecast stock price for interrelated markets by Hassan and Nath [61]. The HMM with two states was used to predict regimes in market turbulence, inflation and industrial production index by Kritzman, Page, and Turkington [79]. Nguyen, N. and Nguyen, D. [105] used the HMM for single observation data to predict regimes of some economic indicators and made stock selections based on the performances of these stocks during the predicted regimes. Nguyen [103] used the HMM with both single and multiple observations to forecast economic regimes and stock prices of some technology stocks. Nguyen and her colleague also used the HMM for modeling interest rates [106] and mortgage backed securities [104]. The HMM was also used to analyze stock market trends in [78]. Previously, Lajos [82] applied the HMM to predict S&P daily prices and developed a strategy to trade stock. Before them, Idvall and Jonsson [67] implemented the HMM algorithms on foreign exchange data.

A collection of applications of the HMM for finance were presented in by Mamon and Elliott [42]. However, many of the papers included used the regime-switching model introduced by Hamilton [59]. Based on the principal concepts of state sequences, Hamilton developed a regime-switching model for nonstationary time series and the business cycle. In the regime-switching model (RSM), observation variables were generated by an auto regression model, whose parameters were optimized by a discrete Markov chain. Many researchers have used the RSM and claimed that the model is the same as the HMM. Although the RSM and the HMM are both associated with regimes or hidden states, the RSM should be viewed as a regression model with regime-shift variables. Furthermore, the HMM is a abroad model that allows a more flexible relationship between observation data and its hidden state sequence.

The HMM has been used for local or national stock trading but not for the global stock market. The biggest global stock market is the All Country World Index (ACWI). The ACWI recently has more than 2,600 stocks issued by 47 countries all over the world. In this study, we will use the HMM to select global stocks for our portfolio. The paper is organized as follows: Section 4.7.2 gives a brief introduction of the ACWI and Section 4.7.3 presents chosen global economics indicators. Section 4.7.4 presents five stock factors that effectively represent performances of the stock. Section 4.7.5 describes how to use the HMM to select global stock portfolios. Section 4.7.6 compares the global stock trading results of using the HMM with not using the HMM, and Section 4.7.7 gives conclusions on our findings.

4.7.2 All Country World Index

The ACWI is developed and maintained by Morgan Stanley Capital International (MSCI). The ACWI consists of large and mid cap representation across 23 Developed Markets and 24 Emerging Markets countries. In 2017, the index had about 2,490 constituents (or stocks) which cover approximately 85% of the global investable equity opportunity set. Figure 4.43 presents distribution of AWCI by physical locations of companies that have stocks in the ACWI. The largest component of AWCI—nearly one-quarter—is the U.S. stock market, followed by Europe and Asia, each representing approximately one-fifth of the global stock universe. There are many stocks that are trading in the U.S. market but issued by companies located out side of the U.S., which makes the U.S. the biggest stock market in the world. Figure 4.44 is the chart of the distribution of the ACWI by stock market cap in August 2017.

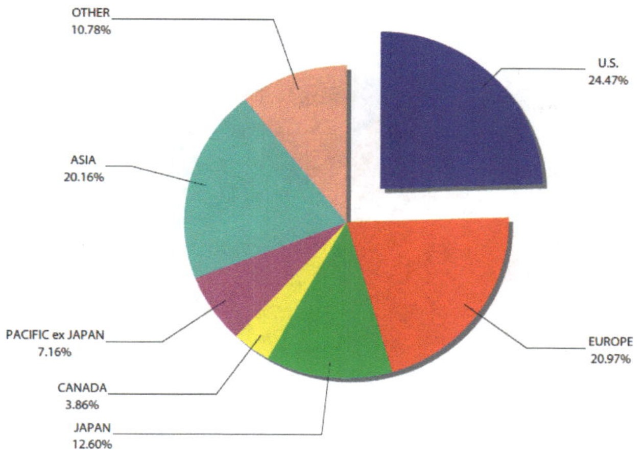

FIGURE 4.43
Distribution of ACWI, August 2017, by financial Companies Locations.

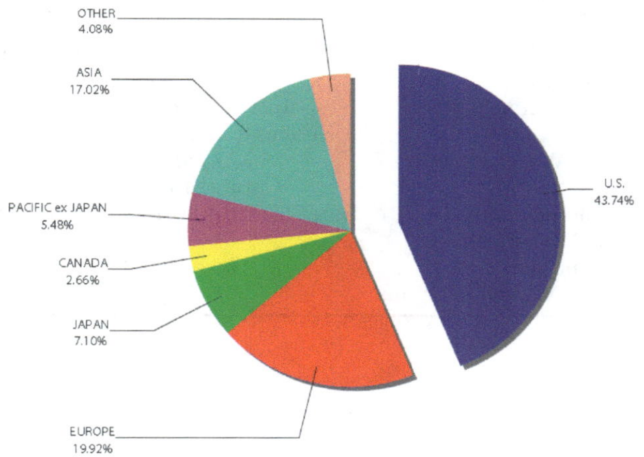

FIGURE 4.44
Distribution of ACWI, August 2017, by market caps.

In this chart, the U.S. stock market consumes almost half of the ACWI, followed by Japan, the United Kingdom, China, and France, with each of these countries allocating less than ten percent of the market cap. Stocks that have big market caps are Apple, Microsoft, Amazon, MSCI India ETF, and Facebook. Figure 4.45 displays the distribution of the ACWI by sectors. The financial sector is the leading sector in the ACWI, followed by the industrial sector. The smallest factor is the telecommunication services' sector. The number of ACWI stocks is subject to change every year as new stocks will be issued.

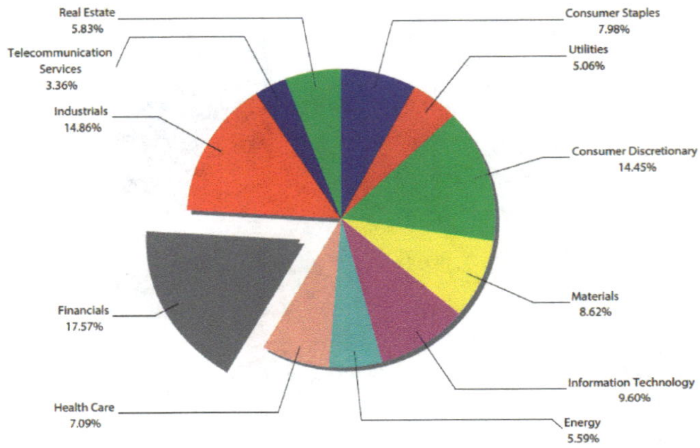

FIGURE 4.45
Distributions of ACWI, August 2017, by sectors.

TABLE 4.10
World Economic Indicators and their sources

Indicator	Title	Source	Data Used
Inflation	Consumer Price Index	International Monetary Fund	Yearly Change
Production	Industrial Production	Haver Analytic	Monthly Change
Sentiment	Economic Climate Index	Institute for Economic Research University of Munich	Original series
Market	MSCI World Index	Morgan Stanley Capital International	Monthly Change
Debt	Public and Private Non-Financial Debt	Bank for International Settlements	Monthly Change
Inflation Expectation	Manufacturing Purchasing Managers' Index	Haver Analytic	Yearly Change

4.7.3 Global Economic Indicators

We choose six main global macroeconomic variables that have significant impacts on the global stock market: inflation, production, sentiment, market, debt, and inflation expectation. The frequency of these data is monthly, and we use their monthly changes, or yearly changes, as input data for our model. The list of these indicators and their sources are given in Table 4.10. We plot the correlation coefficient between the indicators in Table 4.11. The correlation coefficients of the global economic indicators in Table 4.11, are weak, or moderate, which indicate that the indicators are independent in representing the economic universe.

4.7.4 Selected Stock Factors

Stock factor investing is gaining more popularity. However, evidence shows that the factor performs significantly different in various macroeconomic regimes. This section aims at selecting the best stock factors given the current economic environment, and as such, improving our portfolio performances. Investors have been analyzed in many stock criteria to choose the most important factors. Graham [56] suggested these three factors that affect the stock

TABLE 4.11
Correlation Coefficient of the World Economic Indicators

Indicator	Inflation	Production	Sentiment	Market	Debt	Inflation Expectation
Inflation	1.00					
Production	0.15	1.00				
Sentiment	0.28	0.07	1.00			
Market	0.15	0.02	-0.04	1.00		
Debt	0.35	0.03	-0.58	0.19	1.00	
Inflation Expectation	0.64	0.17	0.13	0.17	0.58	1.00

market performances: the real growth, inflationary growth, and speculative growth. Researchers at Ned Davis Research Group [121] examined about two hundred factors and suggested the five most powerful stock factors: Free Cash Flow/Enterprise Value, Shareholder Yield, Lower Accruals Ratio, Operating Cash Flow/Assets, and Price Momentum. Based on these suggestions and our own analysis of the ACWI, we chose five stock factors for the world stock selection: Free Cash Flow/Enterprise Value, Earnings/Price, Sales/Enterprise, Long-term Sales Growth (LTSG), and Long-term Earning Per Share Growth (LTEPSG). Below are definitions of the factors:

1. Free Cash Flow/Enterprise Value is operating cash flow, minus the sum of cash dividends plus capital expenditures, divided by the market value of equity, plus debt.
2. Earning/Price is calculated by the earning accumulation over the trailing twelve months of the stock divided by the weekend price. A higher number indicates a greater value for each unit of earnings, which tends to drive a higher stock return.
3. Sale/Enterprise is calculated by the sales accumulation over the trailing twelve months of the stock divided by the market value of equity, plus debt (enterprise value). A higher sale/enterprise value signifies that each unit of a stock's value is used to generate more sales, which normally leads to a higher stock return.
4. Long-term Sales Growth is the projected long-term growth rate of sales based on a five-year moving regression trend line. A high sale growth rate normally leads to a higher future returns.
5. Long-term Earning Per Share Growth is the projected long-term growth rate of earning per share based on a five-year moving regression trend line. A high earning growth rate normally leads to a higher future returns.

4.7.5 Global Stock Selection Using an HMM

In this section, we will introduce a new global stock selection method using one of the unsupervised machine learning methods, the HMM. Empirical evidences showed that stock factor returns are significantly different in different economics regimes [121]. Therefore, in the stock selection process, we will use the HMM to find hidden states of the economic indicators, and then assign a weight for each stock factor listed in Section 4.7.4. We then calculate the weighted score of each stock to make a stock selection. The stock selection process can be broken down into the following steps. In the first step, we assign scores for each stock factor for all ACWI stocks in the current month. In the second step, we use the unsupervised machine learning, the HMM, to find regimes of the six global economics indicators. The third step is to analyze stock performances in the economics regimes to assign the weight of each of the five stock factors, then calculate the weighted score for each stock. The last step is to choose the top 10% of the global stocks that have the highest weighted score for the trading portfolio. We will introduce basic elements of the HMM and explain the steps of using the HMM to assign a weight for each stock factor in the following sections.

4.7.5.1 Calculating Stock Factor Scores

In the first step, at the end of each month, we calculate five factors of each stock in the global stock pool then convert the values to a score from one to one hundred, for each stock factor. Each stock will have 5 factor scores, with the range from one to one hundred. For example, at the end of each month, we calculate the Earning/Price factor of all stocks in the ACWI, rank them from one to to the total number of stocks, (the number of stocks in ACWI is subject to change each month), divide this rank by the number of stocks and multiples with one hundred to have the score of the stock factor. Table 4.12 is an example of a table of stock factor scores and their equal weighted score. In this table, the EPS is the earning per share and the average score, or equal weighted score, of each stock is the average of all of its the five factor scores.

TABLE 4.12
Example of ACWI Stock Factor Scores

Stock	Free Cash Flow Enterprise	Earning Price	Sale Enterprise	Long-term Sale Growth	Long-term EPS	Average Score
A1	80	57	78	89	45	69.8
A1	100	77	70	82	78	81.4
A1	50	73	67	73	55	63.6

We then choose stocks based on the average of its factor scores. However, each stock factor will have a different effect to the stock price. Therefore, using the equal weighted factor scores to select stocks may not be the best choice. We will propose a new method to calculate the weight of each stock factor to measure the composition score of the five stock factors to make a monthly stock selection in the next section. We will use the HMM to find regimes of the global economics indicators then analyze the performances of stock factors in the regimes to assign a weighted scores for each stock.

4.7.5.2 Identifying Regimes in Global Economic Indicators

In the second step, we will use the HMM to find regimes of the six global economics indicators. Note that we apply the HMM on monthly changes of Production, Market, and Debt, yearly changes of Inflation and Inflation Expectation. For the Sentiment data, we use original data. The monthly changes or yearly changes, R_t, are calculated by

$$R_t = \frac{O_t - O_{t-i}}{O_{t-i}},$$

where O_t is observation data at time t, $i = 1$ for monthly changes, and $i = 12$ for yearly changes. We use these returns of the economics indicators to avoid dealing with different units of the data. Figure 4.46 displays performances of the six global economic indicators from December 2004 to September 2017.

We assume that the economics has two regimes which represent a low or high of economics's returns. So, in this study, we use the HMM with two regimes, where regime one represents a low economics return period and regime two represents a high economics return period. We also assume that the observation probability of each indicator on a given regime

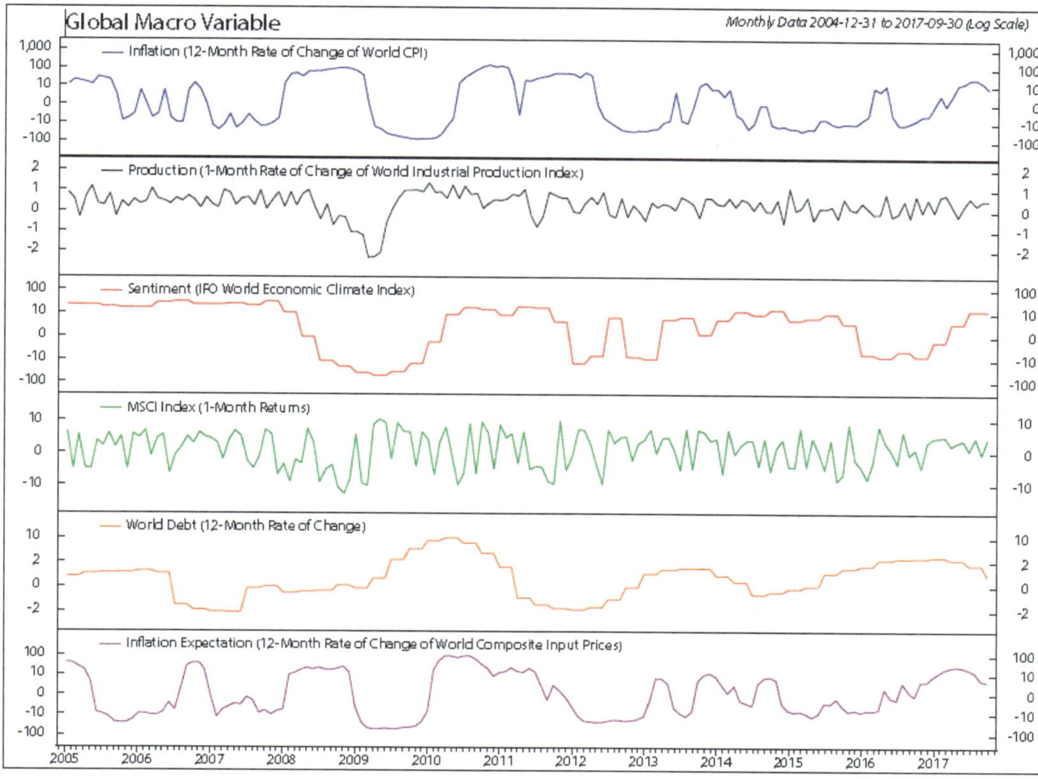

FIGURE 4.46
Performances of six global economics indicators from 2004 to 2017.

follows a normal distribution with mean μ and standard deviation σ. At the end of each month, we use the Baum-Welch algorithm of the HMM to find its parameters

$$\lambda = \{\mathbf{A}, \mu, \sigma, \mathbf{p}\},$$

where $\mu = (\mu_1, \mu_2)$, $\sigma = (\sigma_1.\sigma_2)$, $\mathbf{p} = (p_1, p_2)$, and $\mathbf{A}$ is a two by two transitional probability matrix. We then calculate the ratios, $\frac{\mu_1}{\sigma_1}$ and $\frac{\mu_2}{\sigma_2}$. We define regime one as the regime with the lower ratio, and regime two as the regime with a higher ratio. In reality, the regime one of an indicator is the regime in which the indicator has a lower mean and higher standard deviation to that of regime two. In a special case when the regime has a lower mean but also a lower standard deviation compared with the other regime, the regime one will have a lower ratio of the mean to its standard deviation. In Figures 4.47-4.60 we present the six global economics indicators with their corresponding regimes using monthly data from December 1999 to August 2017. In each of the figures, below the graph are two tables. The table one on the left-hand side presents the means and standard deviations of the two regimes while the table on the right-hand side displays the mean and standard deviations of the returns of the observation data. We can see from all of these tables that regime one has lower mean and higher standard deviation. The data analysis tables of the real data in each regime give the consistency results with the regimes' variable results. The results from the Figures 4.47-4.60 indicate that each global economics indicator has its regimes in different subperiods.

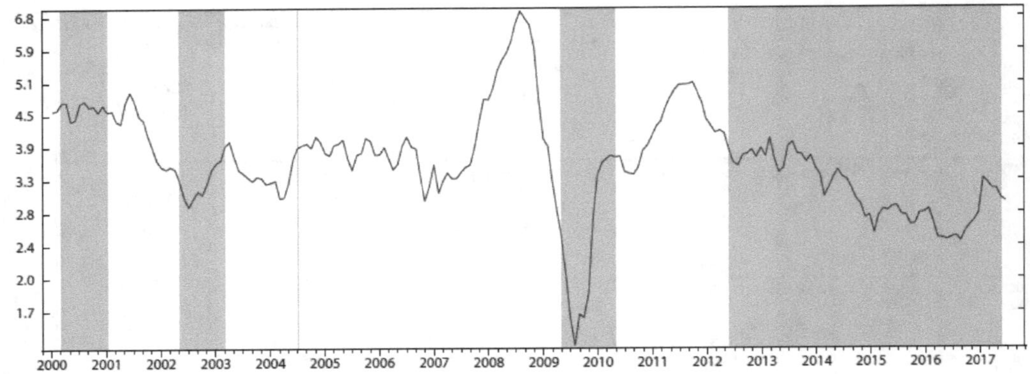

Regime Parameters

Mu		Sigma	
Regime 1 (Shaded)	Regime 2 (Unshaded)	Regime 1 (Shaded)	Regime 2 (Unshaded)
-10.30	9.15	13.21	48.59

Data Statistics

Mean		Variance	
Deflation (Shaded)	Inflation (Unshaded)	Deflation (Shaded)	Inflation (Unshaded)
-13.85	14.89	500.22	1,302.68

FIGURE 4.47
The world inflation and its regimes, monthly data from 12/31/1999 to 08/31/2017.

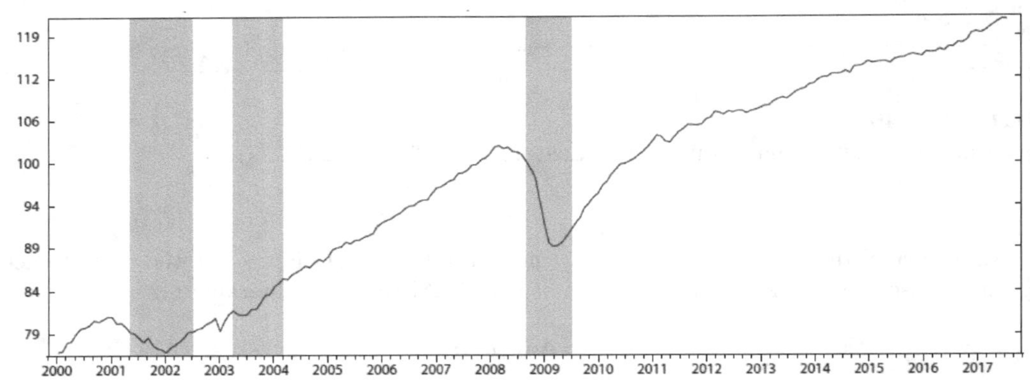

Regime Parameters

Mu		Sigma	
Regime 1 (Shaded)	Regime 2 (Unshaded)	Regime 1 (Shaded)	Regime 2 (Unshaded)
-0.68	0.31	0.99	0.38

Data Statistics

Mean		Variance	
Contraction (Shaded)	Expansion (Unshaded)	Contraction (Shaded)	Expansion (Unshaded)
-0.16	0.29	1.05	0.17

FIGURE 4.48
The world production and its regimes, monthly data from 12/31/1999 to 08/31/2017.

However, they all are in regime one, except the World Debt in Figure 4.51 during the global economic crisis in 2008-2009.

In Figure 4.47, regime one represents the deflation period and regime two presents the inflation period. In the deflation period, we have a negative yearly change and a higher variance compared with yearly changes of the data in the inflation period.

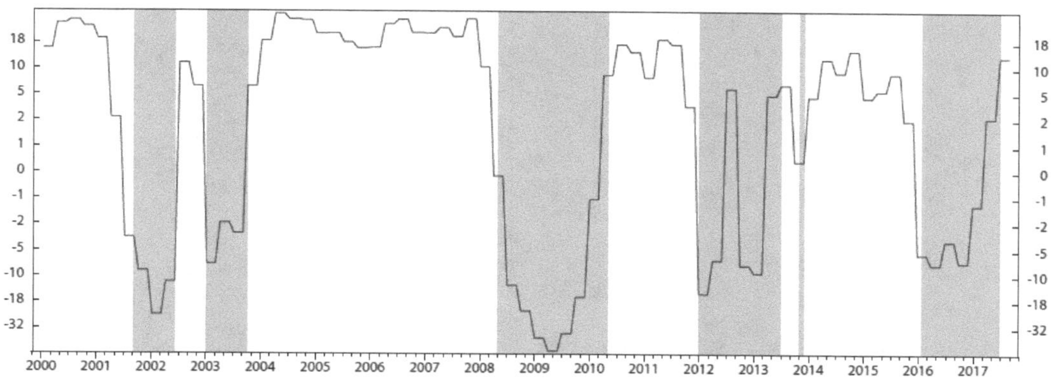

Regime Parameters

Mu		Sigma	
Regime 1 (Shaded)	Regime 2 (Unshaded)	Regime 1 (Shaded)	Regime 2 (Unshaded)
-10.80	13.91	11.69	8.53

Data Statistics

Mean		Variance	
Depressed (Shaded)	Elevated (Unshaded)	Depressed (Shaded)	Elevated (Unshaded)
-9.64	15.93	197.11	88.15

FIGURE 4.49
The world sentiment and its regimes, monthly data from 12/31/1999 to 08/31/2017.

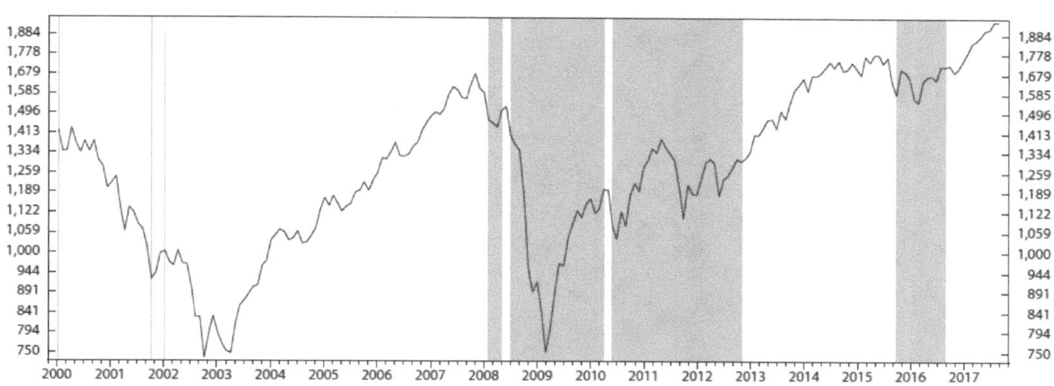

Regime Parameters

Mu		Sigma	
Regime 1 (Shaded)	Regime 2 (Unshaded)	Regime 1 (Shaded)	Regime 2 (Unshaded)
-0.23	1.15	5.42	2.49

Data Statistics

Mean		Variance	
Bear (Shaded)	Bull (Unshaded)	Bear (Shaded)	Bull (Unshaded)
-0.07	0.46	35.13	11.61

FIGURE 4.50
The MSCI world index and its regimes, monthly data from 1999/12/31 to 2017/08/31.

In Figure 4.48, regime one represents the contraction period and regime two presents the expansion period of the World Production. From December 2000 to August 2017, the World Production had only three contraction periods. During the contraction periods, the World Production has mean and variance of the monthly changes as −0.16 and 1.05 respectively. The regime one corresponding to the contraction periods has it's mean and standard deviation of −0.68 and 0.099, respectively. In contrast to the contraction periods, during

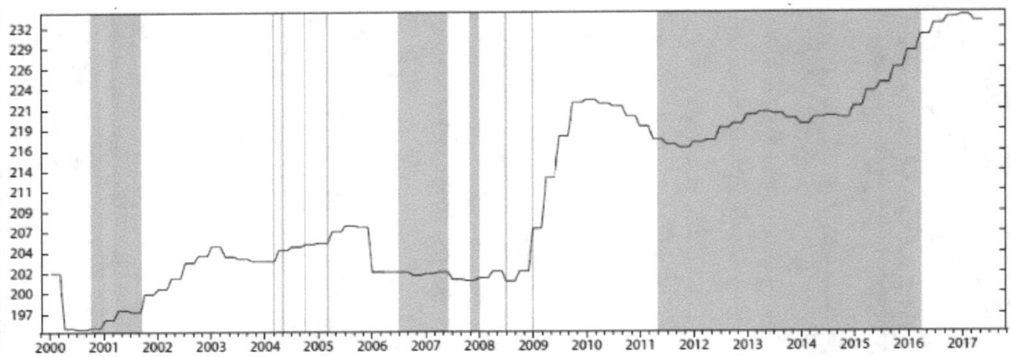

Regime Parameters

Mu		Sigma	
Regime 1 (Shaded)	Regime 2 (Unshaded)	Regime 1 (Shaded)	Regime 2 (Unshaded)
-0.32	3.51	1.24	2.35

Data Statistics

Mean		Variance	
Low (Shaded)	High (Unshaded)	Low (Shaded)	High (Unshaded)
0.11	1.50	2.68	7.16

FIGURE 4.51
The world debt and its regimes, monthly data from 12/31/1999 to 08/31/2017.

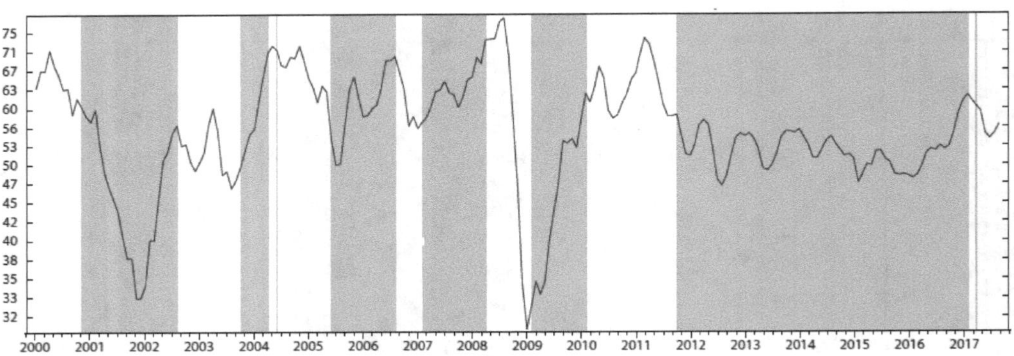

Regime Parameters

Mu		Sigma	
Regime 1 (Shaded)	Regime 2 (Unshaded)	Regime 1 (Shaded)	Regime 2 (Unshaded)
-9.88	36.86	16.22	25.43

Data Statistics

Mean		Variance	
Low (Shaded)	High (Unshaded)	Low (Shaded)	High (Unshaded)
-3.86	21.21	530.69	946.19

FIGURE 4.52
Global inflation expectations and their regimes, based on monthly data from December 1999 to August 2017.

the expansion periods the World Production had a positive mean and a smaller variance of monthly changes.

In Figure 4.49, regime one displays the depressed period and regime two presents the elevated period of the World Sentiment. The HMM captures the depressed regimes of the World Sentiment five times during the period from 1999 to 2017, in Figure 4.49. The data analysis table at the bottom of Figure 4.49 shows that in regime one (which presents the

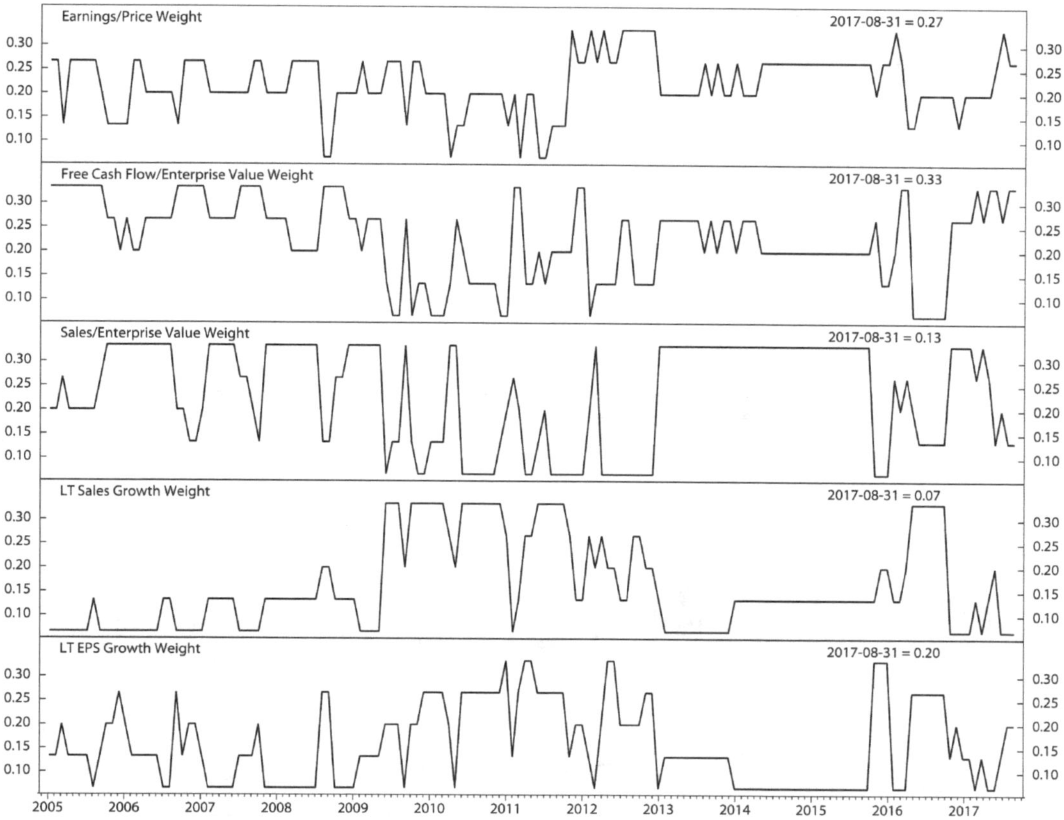

FIGURE 4.53
Calculated weights of five global stock factors from January 2005 to September 2017.

depressed state of the World Sentiment), the observation data was assumed to come from a normal distribution with the mean of −10.80 and standard deviation of 11.69. The mean of the World Sentiment in the depressed periods is −0.9.6, and the standard deviation is 14.04, (the variance is 197.11). While in the elevated periods (presented by the regime two), the World Sentiment is 15.93 and its standard deviation is 9.39.

In Figure 4.50, regime one and regime two represent the Bear and Bull market of the MSCI World Index, respectively. During the Bear market period, the MSCI World Index had a negative mean and a very large variance (35.13) of its monthly changes.

In Figure 4.51, regime one and regime two represent the Low and High periods of the World Debt respectively. In Figure 4.60, regime one and regime two represent the Low and High periods of the World Inflation Expectation, respectively.

After each month, we update data and re-calibrate model parameters, predicting the economics indicators' regimes, and keep moving on. Thus, we will not have the same regimes of economics indicators each month.

4.7.5.3 Assigning Stock Composition Scores

In order to assign a weight for a stock factor, we will analyze its performances in a similar economic condition. In Section 4.7.5.2, we used the HMM to discover the hidden regimes of

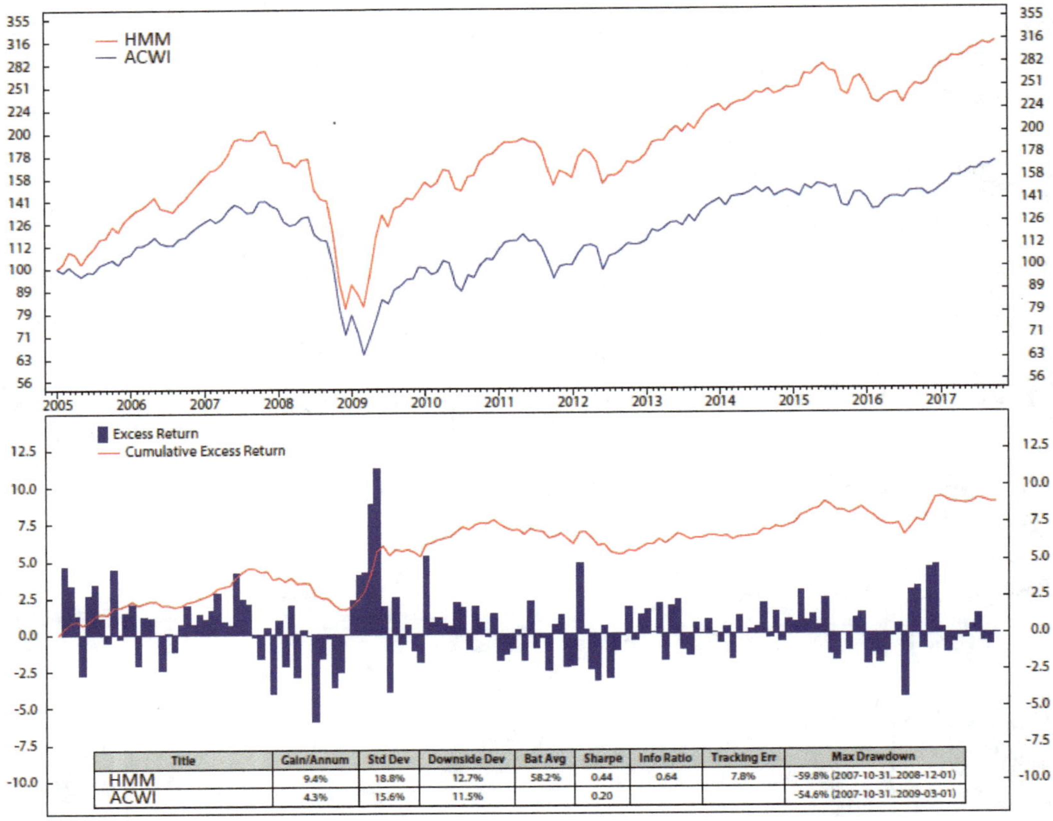

FIGURE 4.54
Monthly performance of the global stock selection portfolio using an HMM, compared to the benchmark ACWI, from January 2005 to September 2017 (log scale).

economic indicators, then assigned a weight for each stock factor based on its performances on each regime to make a composition score of its five factors. Based on this weighted score, we will make a monthly selection of 10% of the ACWI for our investment portfolio. In the second step, we used the HMM to find regimes of the six global economic indicators: inflation, production, sentiment, market, debt, and inflation expectation, that were defined in Section 4.7.3, using historical data up to the recent month. Then, we predicted regimes of the indicators in the next month. After predicting the regimes of the six global economic indicators for the next month, we looked back at historical data (back to December 1999) and found similar regimes of these six variables. For example, if the predicted regimes for the next month of these (inflation, production, sentiment, market, debt, and inflation expectation) are $(1, 1, 2, 2, 1, 1)$ respectively, we will look from recent time back in historical data to find the months that these six indicators had the same regimes $(1, 1, 2, 2, 1, 1)$. We then checked the performance of each stock factor (defined in Section 4.7.4) in the period of three months after each similar set of regimes. Based on these performances, we rank the factors from one to five. The weight of each factor is calculated by taking its rank divided by the sum of 15, the sum of all the possible ranks. We present the weights of the five stock factors in Figure 4.53.

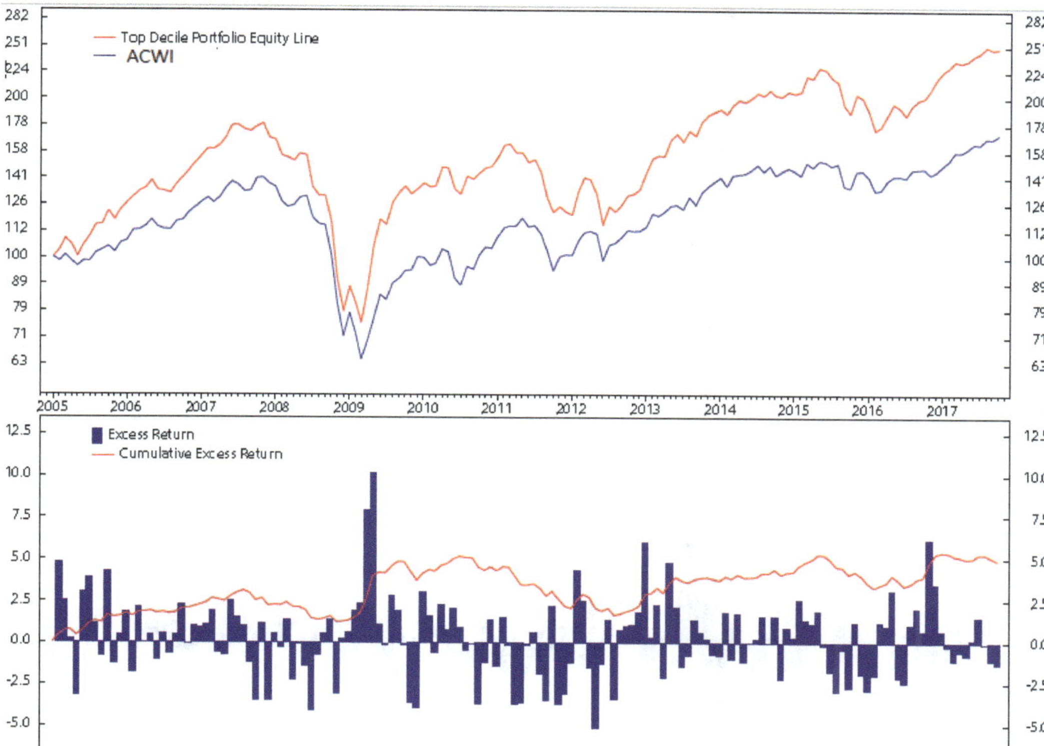

Title	Gain/Annum	Std Dev	Downside Dev	Bat Avg	Sharpe	Info Ratio	Tracking Err	Max Drawdown
Top Decile Portfolio Equity Line	7.4%	18.6%	12.6%	54.2%	0.34	0.39	8.0%	-57.9% (2007-10-31..2009-03-01)
ACWI	4.3%	15.6%	11.5%		0.20			-54.6% (2007-10-31..2009-03-01)

FIGURE 4.55
Global stock selection from January 2005 to September 2017 based on Stock Free Cash Flow/Enterprise Factor.

The composite score for each stock is calculated by the summation of the products of its factor weight with the corresponding score. The final composite scores will have the range from one to one hundred. We select 10% of the ACWI, about 250 stocks in 2017, with the highest composite score for our portfolio. To make a stock selection for the next month and so on, we update data and repeat the selection process. Note that each month, we will calibrate new model parameters, predict new economics regimes, and make a new stock selection. We will sell stocks that are no longer in our list and buy the newly selected stocks.

4.7.6 Results

Figure 4.54 shows the monthly performances of our global stock portfolio compared to the performance of the whole ACWI, from January 2005 to September 2017. In Figure 4.54, we compare our global selected stock portfolio using the machine learning HMM with the ACWI. We scale the stock prices so that the price at the beginning of the stock selection process (January 2005) could be $100.00.

At the bottom of the Figure 4.54 is a table that summarizes the performances. In the table, we present the annual gain (gain/annum), standard deviation, downside deviation,

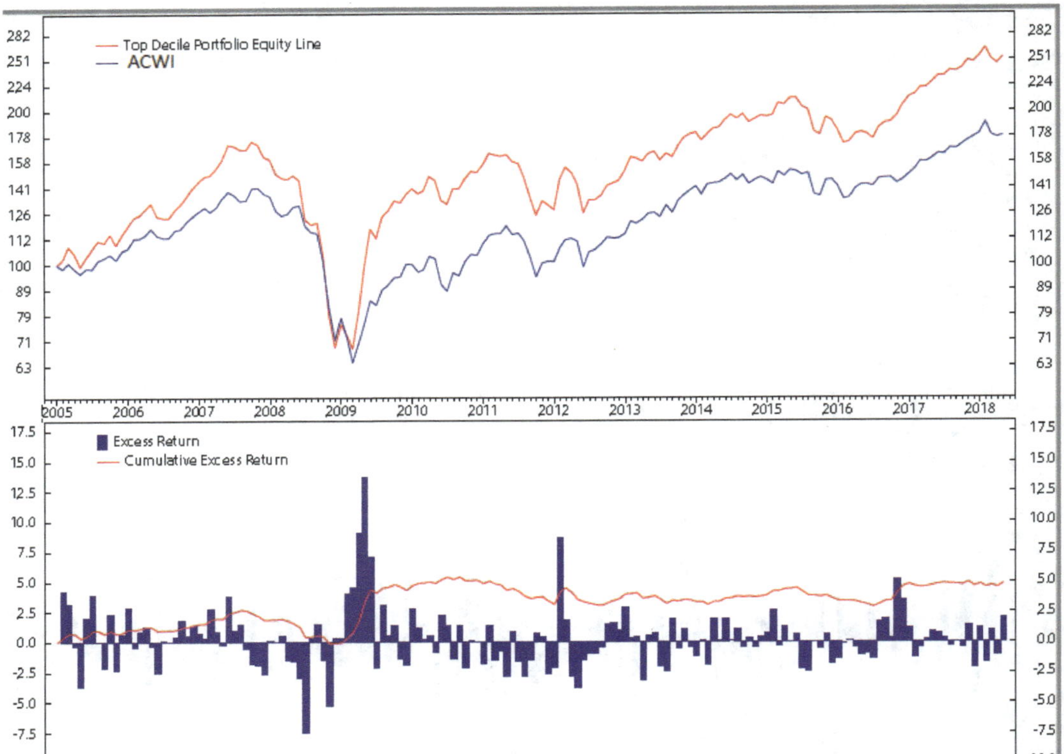

Title	Gain/Annum	Std Dev	Downside Dev	Bat Avg	Sharpe	Info Ratio	Tracking Err	Max Drawdown
Top Decile Portfolio Equity Line	7.2%	19.2%	12.8%	56.3%	0.31	0.33	8.4%	-60.8% (2007-09-30..2009-03-01)
ACWI	4.4%	15.4%	11.3%		0.21			-54.6% (2007-10-31..2009-03-01)

FIGURE 4.56
Global stock selection from January 2005 to September 2017 based on Stock Earning/Price Factor.

batting average, Sharpe ratio, information ratio (info ratio), tracking error, and the maximum drawdown of the data. The batting average is a statistical measure used to measure our investment model's ability to meet or beat the benchmark ACWI. Recall that the batting average is calculated by dividing the number of months in which the single factor portfolio beats or matches the ACWI by the total number of months in the period of question, and multiplying that factor by 100%. In other words, the batting average is the percent of the time the single factor portfolio returns meet/exceed the ACWI returns. The tracking error measures the consistency of excess returns. It is created by taking the difference between the single factor portfolio return and the benchmark ACWI return every month, and then calculating how volatile that difference is. Tracking error is also useful in determining just how "active" a model's strategy is. The lower the tracking error, the closer the model follows the benchmark. The higher the tracking error, the more the model deviates from the benchmark. In this study, the Sharpe ratio is calculated by taking the ratio of excess returns and its standard deviation. The results from Figure 4.54 show that with an initial investment of \$100.00 in January 2005, we gain \$316.00 in September 2017 using the HMM, compared to the gain of \$170.00 of the All Country World Index, ACWI. The results in the table bellow the graph of the Figure 4.54 also indicate that our HMM portfolio outperformed the benchmark ACWI.

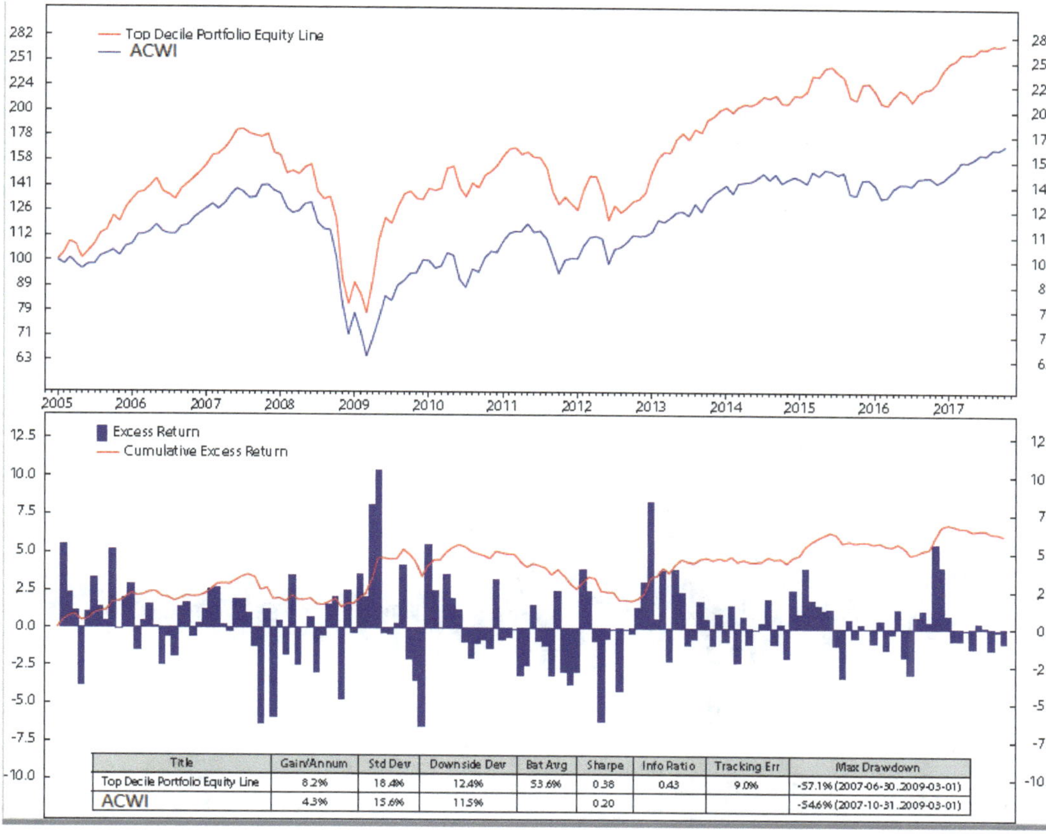

Title	Gain/Annum	Std Dev	Downside Dev	Bat Avg	Sharpe	Info Ratio	Tracking Err	Max Drawdown
Top Decile Portfolio Equity Line	8.2%	18.4%	12.4%	53.6%	0.38	0.43	9.0%	-57.1% (2007-06-30..2009-03-01)
ACWI	4.3%	15.6%	11.5%		0.20			-54.6% (2007-10-31..2009-03-01)

FIGURE 4.57
Global stock selection from January 2005 to September 2017 based on Stock Sales/Enterprise Factor.

We go further by comparing the performance of our portfolio using the HMM with the two common stock selection methods: a single factor method and the equal weighted method. In the single factor method, stocks will be selected based on scores of a single stock factor. In the equal weighted method, stocks will be selected using the average score of the five stock factors, which was explained in Table 4.12. Each month, we choose the top 10% stocks that have the highest scores. For the next month, we update data and repeat the process to make a new portfolio. We sell ones that are not on the elected list while buying the newly-entered ones. Below are the results of the monthly performances of the selected global stock portfolio from January 2005 to September 2017. Figure 4.55 presents our stock selection portfolio (10% of the ACWI) based on the stock's Free Cash Flow/Enterprise Value. The results of selecting global stocks base on the four other factors: Earning/Price, Sale/Enterprise, Long-term Sale Growth, and Long-term Earning Per Share Growth are presented in Figures 4.56-4.59. In each figure, we compare our global selected stock portfolio with the ACWI. We scale the stock prices so that the price at the beginning of the stock selection process (January 2005) could be \$100.00. We also calculate the monthly excess return (in percentage) of our portfolio to the returns of the ACWI benchmark, and the accumulation of the excess returns in each figure. We can see from Figures 4.55-4.59 that our single stock factor portfolio performs better than the ACWI. Figure 4.55 show that

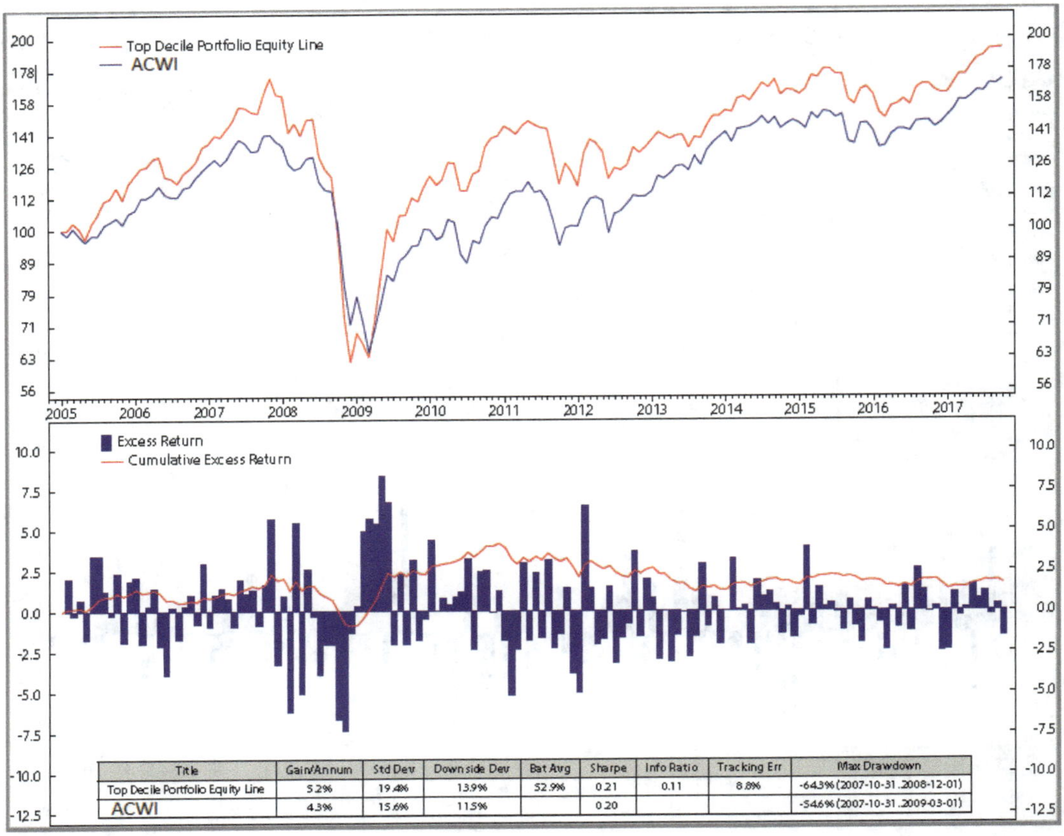

FIGURE 4.58
Global stock selection from January 2005 to September 2017 based on Stock Long-term Sales Growth Factor.

with the initial investment of $100.00 in January 2005, our portfolio based on the ranking of the Stock's Free Cash Flow/Enterprise Value reached a value of $251.00 in September 2017, while the whole ACWI had a value of $178.00. Among the stock factor models, the portfolio based on the factor Sale/Enterprise in Figure 4.57 yields the highest return. The portfolio based on the factor yield the annual return of 9.4% while the ACWI yields 4.3%. It also shows that our portfolio has the cumulative monthly excess return around 6% when compared to the returns of the ACWI. On the other hand, the model based on the Long-term Earning Per Share Growth in Figure 4.59 is the worst model among the five factor models. Its chosen portfolio perform just a little bit better than the ACWI with the excess return of less than 1%.

Figure (4.60 displays the performance of a stock selection portfolio based on the equal weight method, where in each factor has an equal weight of 20%, and the benchmark ACWI. We can see that the equal weight model does not do as well as the single factor model performed in Figures 4.55-4.59. The equal-weighted model yields an annual gain of 4.5% while the ACWI gets an annual gain of 4.3%. However, the equal-weighted portfolio has a higher risk than the ACWI, which leads to a lower Sharpe ratio than the ACWI. Thus, the weighted score method using the HMM outperforms the two common methods: the equal-weighted and the single factor.

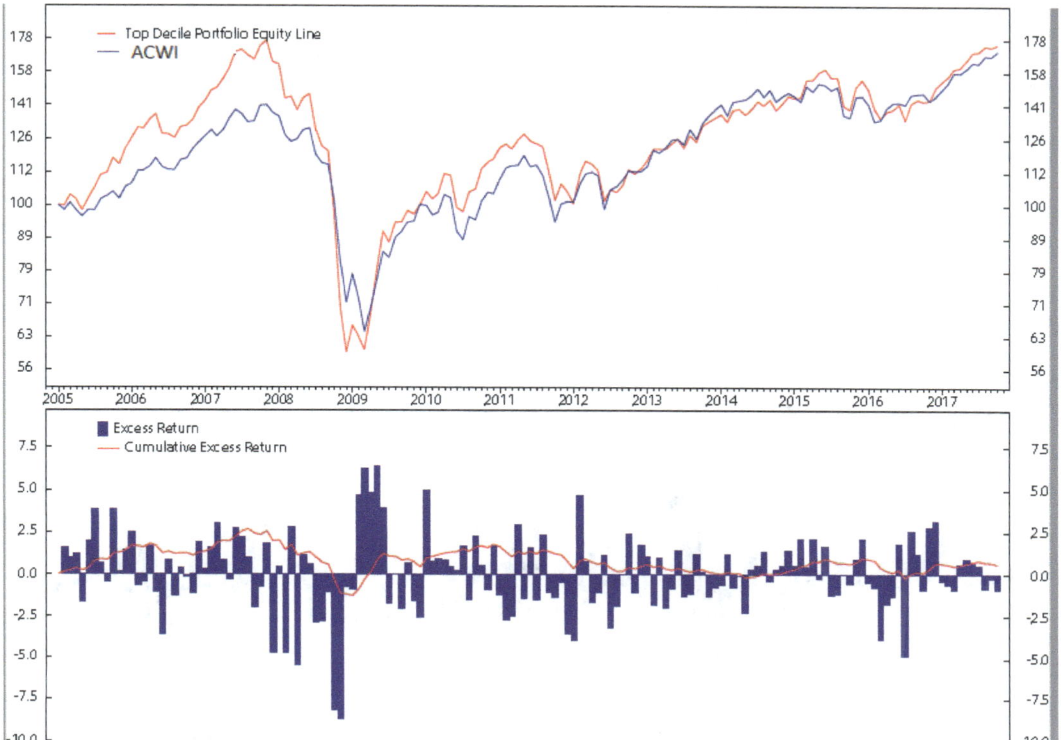

Title	Gain/Annum	Std Dev	Downside Dev	Bat Avg	Sharpe	Info Ratio	Tracking Err	Max Drawdown
Top Decile Portfolio Equity Line	4.5%	18.8%	14.1%	51.6%	0.18	0.02	7.9%	-66.2% (2007-10-31 .2008-12-01)
ACWI	4.3%	15.6%	11.5%		0.20			-54.6% (2007-10-31 .2009-03-01)

FIGURE 4.59
Global stock selection from January 2005 to September 2017 based on Stock Long-term Earning Per Share Growth Factor.

Our weighted model, using the HMM, outperforms the equal-weighted model in Figure 4.60 and all of the single factor models in Figures 4.55-4.59. With the annual return of 9.4%, the excess return of 7.75%, and the Sharpe ratio of 0.44 during the testing period from January 2005 to September 2017. In Figure 4.54, the model stands out from other investigated models in stock trading. The HMM is an efficient unsupervised machine learning model in global stock selections.

4.7.7 Conclusions

We presented the application of the unsupervised machine learning HMM for global stock selections. Each month, we calculated the score for each stock in the global stock market based on the five stock factors. We then used the HMM to predict regimes of the six global economic indicators and analyzed the performances of each stock factor to calculate a weighted score for each stock. Finally, we chose 10% of the global stocks which had the highest weighted scores for our monthly trading portfolio. We compared the returns of our investment with the results of using a single stock factor. The results showed that the stock portfolio selected by using the weighted score of the five stock factors performed better than using one based on a one stock factor. We also compared our portfolio, the

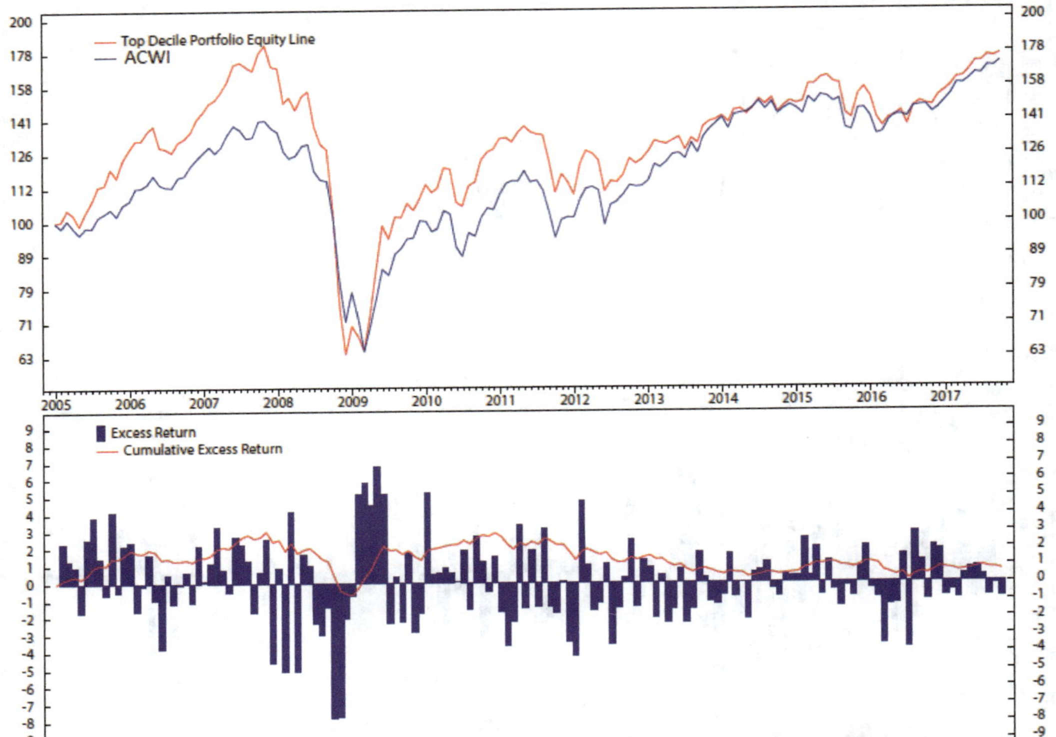

Title	Gain/Annum	Std Dev	Downside Dev	Bat Avg	Sharpe	Info Ratio	Tracking Err	Max Drawdown
Top Decile Portfolio Equity Line	4.5%	19.0%	14.1%	52.3%	0.17	0.02	8.2%	-65.2% (2007-10-31..2008-12-01)
ACWI	4.3%	15.6%	11.5%		0.20			-54.6% (2007-10-31..2009-03-01)

FIGURE 4.60
Equal model, top decile of valuation composite performance, monthly global stock selection from January 2005 to September 2017, (log scale).

HMM weighted factor portfolio, with the portfolio that includes all global stocks and the portfolio selected by using the equal weight of the five stock factors. Results demonstrated that the HMM weighted factor portfolio is the best among the investigated portfolios. The unsupervised machine learning HMM showed that it is an efficient model for global stock portfolio selections.

5

Applications of HMM in Other Disciplines

In this chapter we will summarize applications of the HMM in other fields, such as speech recognition and bioinformatics. Basic steps of applying the HMM to speech recognition will be presented through an example.

5.1 Applications in Speech Recognition

The first application of the HMM was in speech recognition [115] in 1983. From that time, many researchers have been working on applying the model to broader speech data. A review of application of the HMM in speech recognition was published in 2008 [50]. Young and his colleagues from the University of Cambridge provided a free hidden Markov model Toolkit (HTK), written in C program language, for speech recognition [51]. Several people posted their search projects of using the HMM for speech recognition on GitHub with python codes [68, 76, 109, 158]. The content used in this section was based on the cited sources. We use recorded speech data named "audio" created by Håkon Sandsmark [63] to demonstrate the application of the HMM in speech recognition.

The "audio" sound folder includes seven sub-folders which are the names (or labels) of seven fruits: apple, banana, kiwi, lime, orange, peach, and pineapple, each sub-folder has fifteen videos of speech of the corresponding fruit name.

In general, using the HMM for speech recognition involves the following steps:

1. Feature Extraction: the step converts speech signals into a sequence of feature vectors.

   $$Speech\ signals\ ---- > Sequence\ of\ featureO = \{O_1, O_2, ..., O_N\}.$$

   The raw speech signals have a lot of noise, therefore we must extract the noise from the audio signals before using them as input in the HMM (or any other models). The Mel-Frequency Cepstral Coefficients (MFCCs) is the widely used technique for extracting the features from the audio signal. In simpler terms, MFCCs are a set of coefficients that capture the shape of the power spectrum of a sound signal. Using the MFCCs to extract a sound wave into a sequence of features involves the following steps:

   - Frame the signal into short frames (or windows).
   - For each frame, convert the signal from the time domain to the power of spectrum from the frequency domain by applying the Discrete Fourier Transform (DFT)

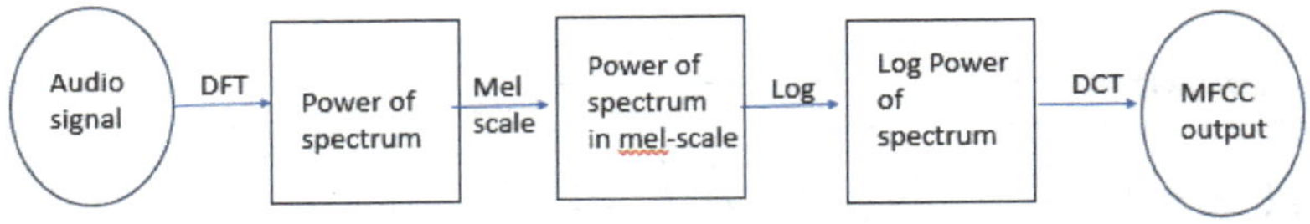

FIGURE 5.1
The sound signals extraction process using the MFCC.

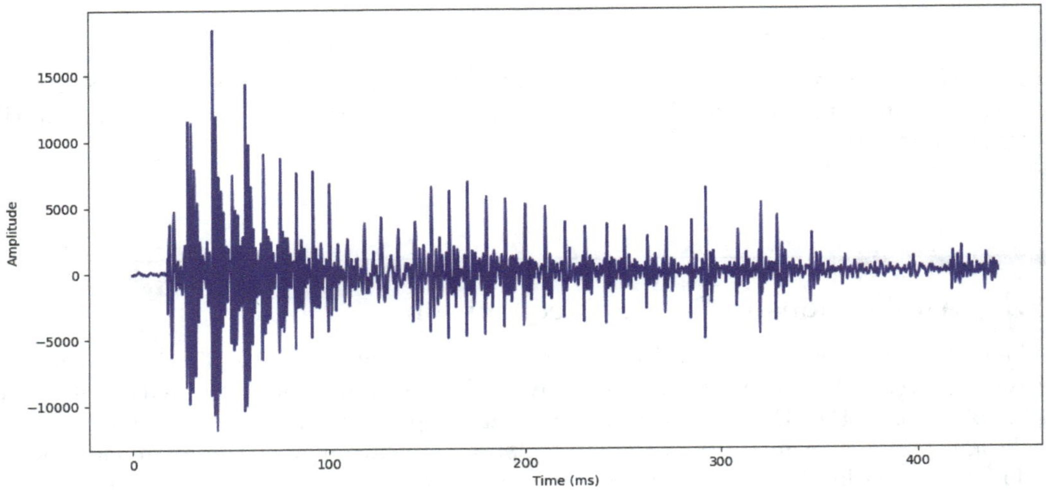

FIGURE 5.2
The signals of an audio sound of the word "banana".

- Map the powers of the spectrum obtained above onto the mel scale by using the formula:
$$M(f) = 1125 * ln(1 + f/700),$$
where f is the frequency and $M(f)$ is the mel scale.
- Take the logarithm of the powers at each of the mel frequencies.
- Take the discrete cosine transform (DCT) of the list of mel log powers.
- The MFCCs are coefficients of the DCT.

Figure 5.1 summarizes the main steps of using the MFCC signal extraction process. Figure 5.2 is the plot of the sound signal of the word "banana" from one file in the "audio" data set. By the sound signal extraction process, the original sound signal in time domain is then transformed to the log of the power spectrum in frequency domain. Figure 5.3 presents the log of the mel-scale power spectrum of the "banana" sound in this step. Finally, Figure 5.4 is the transformation of the wave sound "banana" using the MFCCs. In the figure, the "banana" sound signals were divided into 43 windows, each window has 13 features.

2. Model Training: in this step, the HMM will be trained using the Baum-Welch algorithm. The MFCC data with labels are input of the HMM. After training, the HMM architecture was defined. That means we will obtain the HMM parameter,

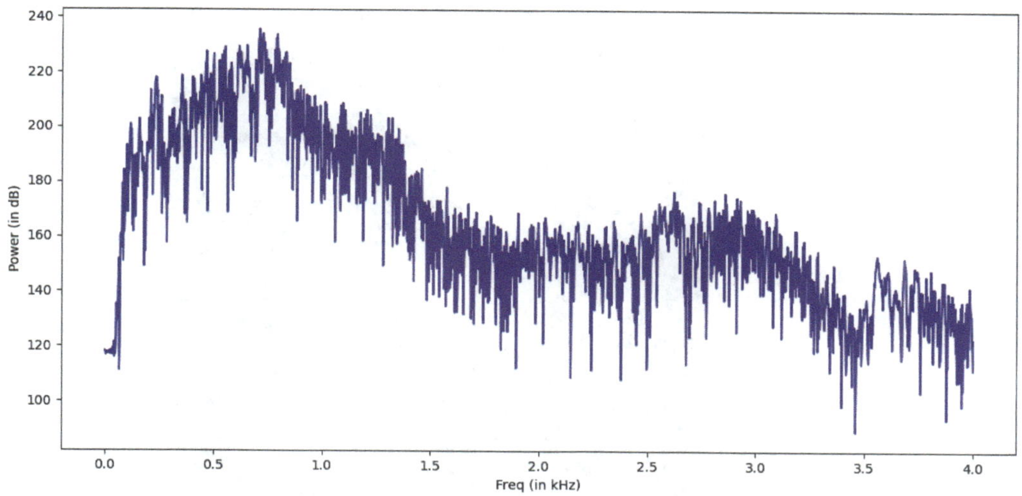

FIGURE 5.3
The log of the power of spectrum in mel-scale of the "banana" sound.

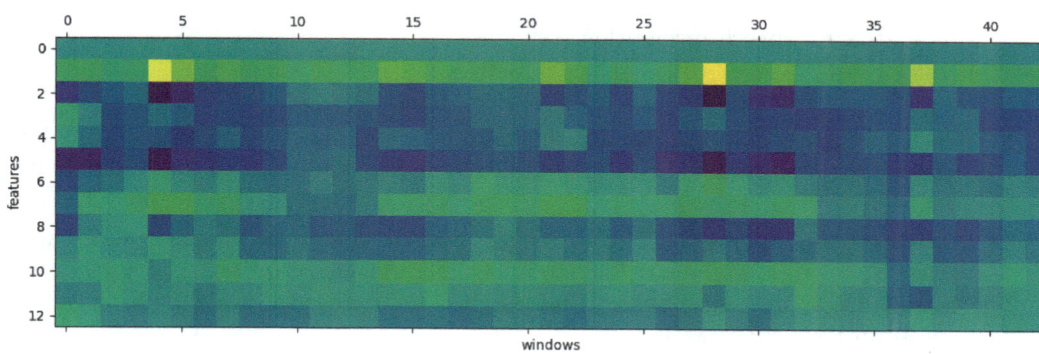

FIGURE 5.4
Sequence of features of the "banana" audio signals after using the MFCCs.

λ, which was defined in Chapter 2. In this "audio" example, we used 80% of the data for training and the remaining 20% for testing.

3. Decoding: in the step, we used the trained HMM to find the most likely hidden state sequences (in this case, the fruit name sequence) of an observed sequence of feature vectors, using the Viterbi algorithm for the testing data. Figure 5.5 is the confusion matrix of the speech prediction for the test data set. The accuracy of this prediction is 80.95%. The result is obtained from a one time run of the code. Since the code used the random number generator for choosing the training and testing set, the accuracy of the prediction will slightly change in each interaction. In many speech recognition systems, an additional language model is used to improve recognition accuracy. This language model can be incorporated into the decoding process to favor sequences of words that are more likely to occur in the given language. Also, for each country language other than English, we may

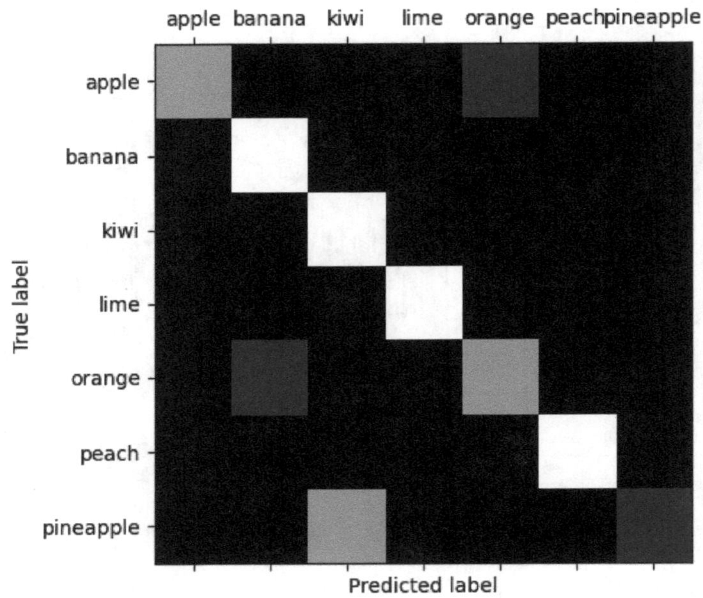

FIGURE 5.5
Confusion matrix for speech recognition.

need additional codes to classify the word or sound labels. The example of the "audio" that we use here is just seven common fruit names, each containing only one word.

4. Post-processing: for a sequence of words, post-processing techniques can be applied. These techniques may include smoothing the transcription, applying grammatical rules, or using contextual information to improve the accuracy of the final transcription.
5. Evaluation: evaluate the performance of the HMM-based speech recognition system using metrics such as word error rate (WER) or phoneme error rate (PER). Adjust parameters and algorithms as necessary to improve the model performance.

   The WER is a common metric used in the evaluation of automatic speech recognition systems. It measures the rate at which the recognized words differ from the original transcript, usually expressed as a percentage. The formula for calculating WER is:

$$WER = \frac{S + D + I}{N},$$

   where: S is the number of substitutions (words that were recognized incorrectly), D is the number of deletions (words in the reference that were not recognized by the system), I is the number of insertions (words that were incorrectly added by the system), and N is the total number of words in the reference transcript. The WER indicates the overall accuracy of the system in converting spoken language into written text, taking into account errors such as missed-recognition, omissions, and additions. A lower WER value indicates higher accuracy, while a higher WER value indicates lower accuracy.

While WER measures the accuracy of entire words, PER focuses specifically on the accuracy of individual phonemes, which are the smallest units of sound in language. The formula for calculating PER is:

$$PER = \frac{S + D + I}{N},$$

where S is the number of substitutions (incorrectly recognized phonemes), D is the number of deletions (phonemes in the reference that were not recognized by the system), I is the number of insertions (incorrectly added phonemes by the system), and N is the total number of phonemes in the reference transcript. Similar to WER, a lower PER value indicates higher accuracy, while a higher PER value indicates lower accuracy in recognizing individual phonemes. PER is particularly useful for evaluating automatic speech recognition systems in tasks where precise phonetic transcription is important, such as in speech synthesis or phonetic research.

Throughout this process, it's important to note that HMMs are just one component of a typical speech recognition system. Other components, such as acoustic models, language models, and pronunciation models, may also be used in conjunction with HMMs to build more accurate and robust speech recognition systems. More details and references on the applications of HMM in speech recognition can be found in [49, 109, 116].

5.2 Applications in Bioinformatics

The HMM has numerous applications in bioinformatics due to its ability to model and analyze sequential data, such as DNA, RNA, and protein sequences. We will present some key applications of the HMM in bioinformatics here.

1. Application in Gene Prediction: HMMs are widely used for gene finding in DNA sequences. They can model the characteristics of coding regions (exons), introns, and intergenic regions, allowing accurate prediction of gene structures in genomes. Krogh [80] used the HMM to predict genes by maximizing the probability of the correct prediction instead of maximizing the probability of the DNA sequence. A package of C^{++} code of the HMM for genes named: GeneMark.hmm was first introduced by Lukashin and Borodovsky [96] in 1998. From that, many researchers have improved upon it for a faster training and more accurate predictions [21, 93, 94, 111, 128]. Data and the package of code can be found at `https://github.com/gatech-genemark/GeneMark-EP-ProtHint-exp`.
2. Application in Sequence Alignment: HMMs are employed in pairwise and multiple sequence alignment algorithms to identify homologous regions and infer evolutionary relationships between sequences [102]. Popular tools include HMMER `https://docs.tinybio.cloud/docs/hmmer-tutorial` and SAMtools `https://github.com/samtools/bcftools/blob/develop/HMM.h` utilize HMMs for sequence alignment tasks.
3. Application in Protein Structure Prediction: HMMs are utilized in protein structure prediction methods to model the probabilistic relationships between amino acids and predict secondary structure elements, such as alpha helices and beta strands. A new method of using the HMM and the Genetic Algorithms to predict

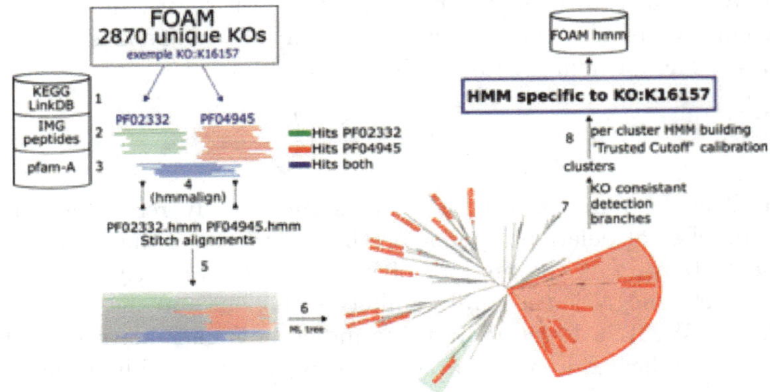

FIGURE 5.6
HMM building pipeline: example with KO:K16157 (methane monooxygenase). Step 1—find Pfam(s) combination assigned to the KO of interest (a) and (b) to check for redundancy. Step 2—fetch IMG peptide sequences which hit the retrieved Pfam(s). Step 3—fetch from Pfam-A database the HMM of interest. Step 4—alignment (hmmalign) and filter each Pfam from extra sequences obtained in IMG. Step 5—stitch filtered alignments. Step 6—draw a Maximum Likelihood tree (fasttree). Step 7—find clusters in tree with same KO. Step 8—split alignment (step 5 output) by cluster (step 7 output) and build the HMM for each, and process the 'Trusted Cutoff' computation. Source: [113].

the secondary structure of proteins was introduced in [147]. The SAM-T08 web server is a protein structure prediction server, which is based on the HMM and is known as the 3D-models, was presented in 2009 [75]. The SAM-T08 web server con be found at `https://openebench.bsc.es/tool/sam-t08`. A faster classification of protein structures was proposed in [99]. In this paper, the authors built a separate HMM for each class of protein structures.

4. Application in Functional Annotation: HMMs are employed for functional annotation of protein sequences by assigning them to specific protein families, domains, or functional sites based on sequence similarity and conserved motifs. A new Functional Ontology Assignments for Metagenomes (FOAM) based on the HMM, was developed [113] to screen environmental metagenomic sequence datasets. The procedure of using FOAM for predicting functional categories was presented in Figure 5.6.

5. Application in Motif Discovery: HMMs can be used to discover over-represented motifs or patterns in DNA or protein sequences, aiding in the identification of transcription factor binding sites, RNA splicing signals, or protein interaction motifs. A review of the HMM and its applications to motif findings can be found in [150]. The HMM was used in the short linear motifs (SLiMs) in proteins to arbitrate a multitude of functions [114]. This method is called the HH-MOTiF method. Adding the HMM to each input sequence of the SLiMs made the HH-MOTiF perform better compared to original SLiMs methods. However, the authors of the paper [114] wrote that: "None of the currently existing SLiM predictors reach an accuracy of more than 35%, including our own method, which again reflects the difficulty of discovering novel SLiMs in proteins and is perhaps inherent to the problem itself". Therefore, there is still a high desire for a better motif prediction model.

We only list a few examples of the diverse applications of the HMM in bioinformatics. The flexibility and probabilistic framework of the model make it a powerful tool for analyzing and interpreting biological sequence data.

5.3 HMM Applications in Other Domains

HMM applications have been expanding to many other areas such as human activity recognition, musicology, and tool wearing. HMMs are frequently used in human activity recognition due to their ability to model sequential data and infer hidden states. The HMM was used in sport analysis by recognizing different types of sports movements or analyzing the performance of athletes. The HMM was first used to recognize tennis player's actions by using time sequential images of a tennis player in [156], in 1992. In 2003, a generic, scale-able and two-layer framework based on the HMM was used for detecting activities of basketball and volleyball players [154]. Many other articles present applications of the HMM in predicting sport activities [6, 10, 74, 101, 119, 144, 151, 152].

HMMs were used to monitor and recognize various human activities, such as walking, running, sitting, standing, and other physical movements in [62, 120, 122, 135]. A continuous-state HMM was used in [139] to detect abnormal activities of multiple moving objects, which were performed by by either cars or a human body. The Hierarchical Dirichlet process HMM, which has an infinite number of states and can automatically decide the optimal number of states, was used with a One-Class Support Vector Machine to find out abnormal activities in [65]. A two-layered HMM architecture was applied for long-term and continuous daily activity monitoring using a wearable body sensor network in [145]. The bottom-layer HMM was used to process sensory data locally at each wireless sensor node and the top-layer HMM was utilized to find the activity sequence from the result of the local processing. A semi-supervised HMM based activity monitoring system was introduced in [155] to not only identify simple events such as sitting, standing and walking, but to also recognizes the behavior or a more complex activity by linking the events together. The method aimed to train a smaller data set compared to a large data set as was required by a general HMM. A two-layered extension of the hidden semi-Markov model (HSMM) for modeling human activities of daily living is presented in [38]. Recently, the Gaussian Mixture HMM was used to recognize human activity using WiFi channel state information [29, 30].

Abnormalities in human daily activities may link to health problems. Thus, the HMM can be applied in healthcare settings to monitor patient activities and detect anomalies, or changes in behavior that may indicate health issues or emergencies. The HMM and the anomalies on sequence patterns of home activities were used to detect the anomalies on sequence patterns of home activities in [112]. According to the paper, the two models gave the same accuracy of predictions. HMMs were used to detect falls or abnormal movements in the elderly or individuals with mobility issues, enabling timely assistance and intervention [70, 89, 130, 131, 133, 134, 160].

The HMM was applied in occupancy sensing by sensing human presence and activities for applications, like energy management, security, and home automation [24, 25, 28, 46, 91, 161]. HMMs can analyze driving patterns and behaviors to detect driver fatigue, distraction, or aggressive driving, contributing to road safety and accident prevention. The HMM applications in driver behavior analysis were presented in [1, 4, 136].

HMMs are used to recognize hand gestures and body movements, which was applied in sign language recognition [3, 7, 52, 83, 142, 143]. The ability to detect gestures, body movements, and speech makes the HMM a great model to enhance human-robot interaction. The

HMM can facilitate human-robot interaction by enabling robots to recognize and understand human activities, gestures, and intentions, enhancing collaboration and communication between humans and robots [34, 92, 110]

In these applications, HMMs are often trained on labeled datasets containing sequences of sensor data corresponding to different human activities. The trained models can then be used to recognize and classify activities in real-time data streams, enabling various practical applications in human activity recognition.

Bibliography

[1] K. Abe, H. Miyatake, and K. Oguri. A study on switching ar-hmm driving behavior model depending on driver's states. In *Proceedings of the IEEE Intelligent Transportation Systems Conference, Bellevue, WA, USA*, pages 806–811. IEEE, 2007.

[2] H. Akaike. A new look at the statistical model identification. *IEEE Transactions on Automatic Control*, 19:716–723, 1974.

[3] M. Al-Rousan, K. Assaleh, and A. Tala'a. Video-based signer-independent arabic sign language recognition using hidden markov models. *Applied Soft Computing*, 9(3):990–999, 2009.

[4] S. B. Amsalu and A. Homaifar. Driver behavior modeling near intersections using hidden markov model based on genetic algorithm. In *Proceedings of the IEEE International Conference on Intelligent Transportation Engineering (ICITE), Singapore*, pages 193–200. IEEE, 2016.

[5] A. Ang and G. Bekaert. International asset allocation with regime shifts. *The Review of Financial Studies*, 15:1137–1187, 2002.

[6] J. Assfalg, M. Bertini, A. Del Bimbo, W. Nunziati, and P. Pala. Soccer highlights detection and recognition using hmms. In *Proceedings of the IEEE International Conference on Multimedia and Expo (ICME)*, volume 1, pages 825–828. IEEE, 2002.

[7] S. G. Azar and H. Seyedarabi. Trajectory-based recognition of dynamic persian sign language using hidden markov model. *Computer Speech & Language*, 61:0885–2308, 2020.

[8] P. M. Baggenstoss. A modified baum-welch algorithm for hidden markov models with multiple observation spaces. *IEEE Transactions on Speech and Audio Processing*, 9(4):411–416, 2001.

[9] A. Bakis. Continuous speech recognition via centisecond acoustic states. *Journal of the Acoustical Society of America*, 59(97), 1976.

[10] F. I. Bashir, A. A. Khokhar, and D. Schonfeld. Object trajectory-based activity classification and recognition using hidden markov models. *IEEE Transactions on Image Processing*, 16(7):1912–1919, 2007.

[11] L. E. Baum and J. A. Egon. An inequality with applications to statistical estimation for probabilistic functions of markov process and to a model for ecology. *Bulletin of the American Meteorological Society*, 73:360–363, 1967.

[12] L. E. Baum and T. Petrie. Statistical inference for probabilistic functions of finite state markov chains. *The Annals of Mathematical Statistics*, 37:1554–1563, 1966.

[13] L. E. Baum, T. Petrie, G. Soules, and N. Weiss. A maximization technique occurring in the statistical analysis of probabilistic functions of markov chains. *The Annals of Mathematical Statistics*, 41:164–171, 1970.

[14] L. E. Baum and G. R. Sell. Growth functions for transformations on manifolds. *Pacific Journal of Mathematics*, 27:211–227, 1968.

[15] A. Ben-Yishai and D. Burshtein. A discriminative training algorithm for hidden markov models. *IEEE Transactions on Speech and Audio Processing*, 12(3):204–217, 2004.

[16] P. Bickel, Y. Ritov, and T. Ryden. Asymptotic normality of the maximum likelihood estimator for general hidden markov models. *The Annals of Statistics*, 26(4):1614–1635, 1998.

[17] P. J. Bickel and Y. Ritov. Inference in hidden markov models i: Local asymptotic normality in the stationary case. *Bernoulli*, 2:199–228, 1996.

[18] B. Bonnie. The dynamic impact of macroeconomic aggregates on housing prices and stock of houses: A national and regional analysis. *Journal of Real Estate Finance and Economics*, 17:1979–1987, 1998.

[19] H. Bozdogan. Model selection and akaike's information criterion (aic): The general theory and its analytical extensions. *Psychometrika*, 52:345–370, 1987.

[20] D. Brigo and F. Mercurio. *Interest Rate Models: Theory and Practice.* Springer, 2nd edition, 2006.

[21] T. Brůna, A. Lomsadze, and M. Borodovsky. Genemark-ep+: Eukaryotic gene prediction with self-training in the space of genes and proteins. *NAR Genomics and Bioinformatics*, 2(2):lqaa026, June 2020.

[22] C. A. Calhoun and Y. Deng. A dynamic analysis of fixed- and adjustable-rate mortgage terminations. *The Journal of Real Estate Finance and Economics*, 24(1):9–33, 2002.

[23] J.Y. Campbell and S.B. Thompson. Predicting the equity premium out of sample: Can anything beat the historical average? *The Review of Financial Studies*, 21(4):1509–1531, 2008.

[24] L. M. Candanedo, V. Feldheim, and D. Deramaix. A methodology based on hidden markov models for occupancy detection and a case study in a low energy residential building. *Energy and Buildings*, 148:327–341, 2017.

[25] E. Chatziantoniou, B. Allen, and V. Velisavljević. An hmm-based spectrum occupancy predictor for energy efficient cognitive radio. In *Proceedings of the IEEE 24th Annual International Symposium on Personal, Indoor, and Mobile Radio Communications (PIMRC), London, UK*, pages 601–605. IEEE, 2013.

[26] C. Chen. How well can we predict currency crises? evidence from a three-regime markov-switching model. Mimeo, 2005.

[27] Z. Chen and R. C. Qiu. Prediction of channel state for cognitive radio using higher-order hidden markov model. In *Proceedings of the IEEE SoutheastCon 2010, Concord, NC, USA*, pages 276–282, 2010.

[28] Z. Chen, Q. Zhu, M. K. Masood, and Y. C. Soh. Environmental sensors-based occupancy estimation in buildings via ihmm-mlr. *IEEE Transactions on Industrial Informatics*, 13(5):2184–2193, 2017.

[29] X. Cheng and B. Huang. Csi-based human continuous activity recognition using gmm–hmm. *IEEE Sensors Journal*, 22(19):18709–18717, 2022.

[30] X. Cheng, B. Huang, and J. Zong. Device-free human activity recognition based on gmm-hmm using channel state information. *IEEE Access*, 9:76592–76601, 2021.

[31] T. Clark and K. West. Approximately normal tests for equal predictive accuracy in nested models. *Journal of Econometrics*, 138(1):291–311, 2007.

[32] J. P. Coelho, T. M. Pinho, and J. Boaventura-Cunha. *Hidden Markov Models Theory and Implementation Using MATLAB®*. CRC Press, 2021.

[33] J.C. Cox, J.E. Ingersoll, and S.A. Ross. A theory of the term structure of interest rates. *Econometrica*, 53(2):385–408, 1985.

[34] Q. Deng and D. Söffker. A review of hmm-based approaches of driving behaviors recognition and prediction. *IEEE Transactions on Intelligent Vehicles*, 7(1):21–31, March 2022.

[35] T. M. T. Do and T. Artieres. Large margin training for hidden markov models with partially observed states. In *Proceedings of the 26th Annual International Conference on Machine Learning (ICML)*, pages 265–272, 2009.

[36] J. Dod. Csi: Credit crunch. `https://www.teru.com/special-report/2007/10/20/csi-credit-crunch`, 2007. The Terui.

[37] J. A. du Preez. *Efficient High-Order Hidden Markov Modelling*. Ph.d. dissertation, University of Stellenbosch, 1997.

[38] T. V. Duong, H. H. Bui, D. Q. Phung, and S. Venkatesh. Activity recognition and abnormality detection with the switching hidden semi-markov model. In *Proceedings of the IEEE Computer Society Conference on Computer Vision and Pattern Recognition (CVPR), San Diego, CA, USA*, volume 1, pages 838–845. IEEE, 2005.

[39] P. Dymarski. *Hidden Markov Models, Theory and Applications*. InTech, Rijeka, Croatia, 2011.

[40] R. J. Elliott, J. B. Moore, and L. Aggoun. *Hidden Markov Models Estimation and Control*. Springer, 1996.

[41] R. J. Elliott and C. A. Wilson. The term structure of interest rates in a hidden markov setting. In R. Mamon and R. J. Elliott, editors, *Hidden Markov Models in Finance*, volume 104 of *International Series in Operations Research & Management Science*, pages 15–30. Springer, 2007.

[42] R. J. Elliott and C. A. Wilson. The term structure of interest rates in a hidden markov setting. In *Hidden Markov Models in Finance*, pages 15–30. Springer, 2007.

[43] H. A. Engelbrecht. *Efficient Decoding of High-Order Hidden Markov Models*. Ph.d. dissertation, University of Stellenbosch, 2007.

[44] C. Erlwein and R. Mamon. An online estimation scheme for a hull-white model with hmm-driven parameters. *Statistical Methods & Applications*, 18(1):87–107, 2009.

[45] C. Erlwein, R. Mamon, and M. Davison. An examination of hmm-based investment strategies for asset allocation. *Applied Stochastic Models in Business and Industry*, 2009.

[46] B. Ai, Z. Fan, and R. X. Gao. Occupancy estimation for smart buildings by an auto-regressive hidden markov model. In *Proceedings of the American Control Conference, Portland, OR, USA*, pages 2234–2239. IEEE, 2014.

[47] G. D. Forney. The viterbi algorithm. *IEEE*, 61:268–278, 1973.

[48] S. Frühwirth-Schnatter. *Finite Mixture and Markov Switching Models.* Springer, 2006.

[49] S. K. Gaikwad, B. W. Gawali, and P. Yannawar. A review on speech recognition technique. *International Journal of Computer Applications*, 10(3):16–24, 2010.

[50] M. Gales and S. Young. The application of hidden markov models in speech recognition. *Foundations and Trends in Signal Processing*, 1(3):195–304, November 2008.

[51] M. J. F. Gales. Cluster adaptive training of hidden markov models. *IEEE Transactions on Speech and Audio Processing*, 8(4):417–428, 2000.

[52] W. Gao, G. Fang, D. Zhao, and Y. Chen. A chinese sign language recognition system based on sofm/srn/hmm. *Pattern Recognition*, 37(12):2389–2402, 2004.

[53] R. Garcia and P. Perron. An analysis of the real interest rate under regime shifts. *The Review of Economics and Statistics*, 78(1):111–125, 1996.

[54] J. Geweke and N. Terui. Bayesian threshold autoregressive models for nonlinear time series. *Journal of Time Series Analysis*, 14:441–454, 1993.

[55] Z. Ghahramani. An introduction to hidden markov models and bayesian networks. *International Journal of Pattern Recognition and Artificial Intelligence*, 15(1):9–42, 2001.

[56] B. Graham. *The Intelligent Investor, Revised Edition.* HarperCollins, New York, 2003.

[57] M. Guidolin and A. Timmermann. Asset allocation under multivariate regime switching. *Journal of Economic Dynamics and Control*, 31:3503–3544, 2007.

[58] M. Guidolin and A. Timmermann. Asset allocation under multivariate regime switching. *Journal of Economic Dynamics and Control*, 31:3503–3544, 2007.

[59] J. D. Hamilton. A new approach to the economic analysis of nonstationary time series and the business cycle. *Econometrica*, 57(2):357–384, 1989.

[60] E. J. Hannan and B. G. Quinn. The determination of the order of an autoregression. *Journal of the Royal Statistical Society: Series B (Methodological)*, 41:190–195, 1979.

[61] M. R. Hassan and B. Nath. Stock market forecasting using hidden markov models: A new approach. In *Proceedings of the IEEE 5th International Conference on Intelligent Systems Design and Applications*, pages 192–196, Wroclaw, Poland, 2005.

[62] J. He, H. Li, and J. Tan. Real-time daily activity classification with wireless sensor networks using hidden markov model. In *Proceedings of the 29th Annual International Conference of the IEEE Engineering in Medicine and Biology Society (EMBC)*, pages 3192–3195. IEEE, 2007.

[63] HMM Speech Recognition Project. Speech data archive. `https://code.google.com/archive/p/hmm-speech-recognition/downloads`. Accessed: 2025-11-03.

[64] T. Honda. Optimal portfolio choice for unobservable and regime-switching mean returns. *Journal of Economic Dynamics & Control*, 28:45–78, 2003.

[65] D. H. Hu, X. X. Zhang, J. Yin, V. W. Zheng, and Q. Yang. Abnormal activity recognition based on hdp-hmm models. In *Proceedings of the 21st International Joint Conference on Artificial Intelligence (IJCAI)*, pages 1715–1720. IJCAI, 2009.

[66] J. Hull and A. White. Pricing interest-rate-derivative securities. *The Review of Financial Studies*, 3(4):573–592, 1990.

[67] P. Idvall and C. Jonsson. Algorithmic trading: Hidden markov models on foreign exchange data. Master's thesis, Linköping University, 2008.

[68] Z. Jackson. Free spoken digit dataset. `https://github.com/Jakobovski/free-spoken-digit-dataset`. Accessed: 2025-11-03.

[69] H. Jiang and X. Li. Incorporating training errors for large margin hmms under semi-definite programming framework. In *Proceedings of the International Conference on Acoustics, Speech, and Signal Processing (ICASSP)*, volume 4, pages 629–632, 2007.

[70] M. Jiang, Y. Chen, Y. Zhao, and A. Cai. A real-time fall detection system based on hmm and rvm. In *Proceedings of Visual Communications and Image Processing (VCIP), Kuching, Malaysia*, pages 1–6. IEEE, 2013.

[71] B. H. Juang. Maximum likelihood estimation for mixture multivariate stochastic observations of markov chains. *AT&T Technical Journal*, 64(6):1235–1249, 1985.

[72] B. H. Juang, S. E. Levinson, and M. M. Sondhi. Maximum likelihood estimation for multivariate mixture observations of markov chains. *IEEE Transactions on Information Theory*, 32(2):307–309, 1986.

[73] B. H. Juang and L. R. Rabiner. Mixture autoregressive hidden markov models for speech signals. *IEEE Transactions on Acoustics, Speech, and Signal Processing*, pages 1404–1413, 1985.

[74] Y. L. Kang, J. H. Lim, M. S. Kankanhalli, C. S. Xu, and Q. Tian. Goal detection in soccer video using audio/visual keywords. In *Proceedings of the IEEE International Conference on Image Processing (ICIP)*, pages 1629–1632. IEEE, 2004.

[75] K. Karplus. Sam-t08, hmm-based protein structure prediction. *Nucleic Acids Research*, 37(2):492–497, 2009.

[76] K. Kastner. Single-speaker word recognition with hidden markov models. `https://github.com/kastnerkyle/kastnerkyle.github.io/blob/master/posts/single-speaker-word-recognition-with-hidden-markov-models/single-speaker-word-recognition-with-hidden-markov-models.ipynb`. Accessed: 2025-11-03.

[77] J. K. Kau, D. C. Keenan, W. J. Muller, and J. F. Epperson. A generalized valuation model for fixed-rate residential mortgages. *Journal of Money, Credit and Banking*, 24(3):279–299, 1992.

[78] G. Kavitha, A. Udhayakumar, and D. Nagarajan. Stock market trend analysis using hidden markov models. arXiv preprint arXiv:1311.4771, 2013.

[79] M. Kritzman, S. Page, and D. Turkington. Regime shifts: Implications for dynamic strategies. *Financial Analysts Journal*, 68:22–39, 2012.

[80] A. Krogh. Two methods for improving performance of an hmm and their application for gene finding. Technical Report 45, Center for Biological Sequence Analysis, 1997. Report No. 4525.

[81] S. A. S. Kumar and V. Ramasubramanian. Automatic language identification using ergodic-hmm. In *Proceedings of the IEEE International Conference on Acoustics, Speech, and Signal Processing (ICASSP '05), Philadelphia, PA, USA*, volume 1, pages I/609–I/612, 2005.

[82] J. Lajos. *Computer Modeling Using Hidden Markov Model Approach Applied to the Financial Markets.* PhD thesis, Oklahoma State University, Stillwater, OK, USA, 2011.

[83] H. K. Lee and J. H. Kim. An hmm-based threshold model approach for gesture recognition. *IEEE Transactions on Pattern Analysis and Machine Intelligence*, 21(10):961–973, 1999.

[84] B. G. Leroux. Maximum-likelihood estimation for hidden markov models. *Stochastic Processes and Their Applications*, 40:127–143, 1992.

[85] B. G. Leroux and M. L. Puterman. Maximum-penalized-likelihood estimation for independent and markov-dependent mixture models. *Biometrics*, 48:545–558, 1992.

[86] S. Levinson, L. Rabiner, and M. Sondhi. An introduction to the application of the theory of probabilistic functions of markov process to automatic speech recognition. *Bell System Technical Journal*, 62:1035–1074, 1983.

[87] J. Li, J. Y. Lee, and L. Liao. A new algorithm to train hidden markov models for biological sequences with partial labels. *BMC Bioinformatics*, 22(162), 2021.

[88] X. Li, M. Parizeau, and R. Plamondon. Training hidden markov models with multiple observations—a combinatorial method. *IEEE Transactions on Pattern Analysis and Machine Intelligence*, 22(4):371–377, 2000.

[89] S. Liang, T. Chu, D. Lin, Y. Ning, H. Li, and G. Zhao. Pre-impact alarm system for fall detection using mems sensors and hmm-based svm classifier. In *Proceedings of the 40th Annual International Conference of the IEEE Engineering in Medicine and Biology Society (EMBC), Honolulu, HI, USA*, pages 4401–4405. IEEE, 2018.

[90] L. A. Liporace. Maximum likelihood estimation for multivariate observations of markov sources. *IEEE Transactions on Information Theory*, 28(5), 1982.

[91] P. Liu, S. K. Nguang, and A. Partridge. Occupancy inference using pyroelectric infrared sensors through hidden markov models. *IEEE Sensors Journal*, 16(4):1062–1068, 2016.

[92] P. Liu, T. Qu, H. Gao, and X. Gong. Driving intention recognition of surrounding vehicles based on a time-sequenced weights hidden markov model for autonomous driving. *Sensors*, 23(21):8761, 2023.

[93] A. Lomsadze, D. P. Burns, and M. Borodovsky. Integration of mapped rna-seq reads into automatic training of eukaryotic gene finding algorithm. *Nucleic Acids Research*, 42(15):e119, September 2014.

[94] A. Lomsadze, V. Ter-Hovhannisyan, Y. Chernoff, and M. Borodovsky. Gene identification in novel eukaryotic genomes by self-training algorithm. *Nucleic Acids Research*, 33(20):6494–6506, November 2005.

[95] H. M. Lu, D. Zeng, and H. Chen. Prospective infectious disease outbreak detection using markov switching models. *IEEE Transactions on Knowledge and Data Engineering*, 22(4):565–577, 2010.

[96] A. V. Lukashin and M. Borodovsky. Genemark.hmm: New solutions for gene finding. *Nucleic Acids Research*, 26(4):1107–1115, February 1998.

[97] R. Mamon and R. J. Elliott. *Hidden Markov Models in Finance*, volume 104 of *International Series in Operations Research & Management Science*. Springer, 2007.

[98] A. A. Markov. Extension of the law of large numbers to dependent quantities (in russian). *Bulletin of the Society of the Physics Mathematics, Kazan, Russia*, 2:135–156, 1906.

[99] G. Mirceva and D. Davchev. Hmm-based approach for classifying protein structures. *International Journal of Bio-Science and Bio-Technology*, 2012.

[100] B. Mor, S. Garhwal, and A. Kumar. A systematic review of hidden markov models and their applications. *Archives of Computational Methods in Engineering*, 28:1429–1448, 2021.

[101] S. Motoi, T. Misu, Y. Nakada, et al. Bayesian event detection for sport games with hidden markov model. *Pattern Analysis and Applications*, 15:59–72, 2012.

[102] D. Mount. Using hidden markov models to align multiple sequences. *Cold Spring Harbor Protocols*, July 2009. PMID: 20147223.

[103] N. Nguyen. *Probabilistic Methods in Estimation and Prediction of Financial Models*. PhD thesis, Florida State University, Tallahassee, FL, USA, 2014.

[104] N. Nguyen. Hidden markov model for portfolio management with mortgage-backed securities exchange-traded fund. Technical report, Society of Actuaries, 2017.

[105] N. Nguyen and D. Nguyen. Hidden markov model for stock selection. *Risks*, 3:455–473, 2015.

[106] N. Nguyen, D. Nguyen, and P. Wakefield. Using the hidden markov model to improve the hull-white model for short rate. *International Journal of Trade, Economics and Finance*, 9(2):54–59, 2018.

[107] O. Nneji, C. Brooks, and C. W. R. Ward. House price dynamics and their reaction to macroeconomic changes. *Economic Modelling*, 32:172–178, May 2013.

[108] B. Nobakht, C. E. Joseph, and B. Loni. Stock market analysis and prediction using hidden markov models. In *Proceedings of the Student Conference on Engineering and Systems (SCES)*, pages 1–4, 2012.

[109] P. Pandey. Isolated speech recognition. `https://github.com/prakashpandey9/IsolatedSpeechRecognition/tree/master`. Accessed: 2025-11-03.

[110] H. S. Park, E. Y. Kim, S. S. Jang, S. H. Park, M. H. Park, and H. J. Kim. Hmm-based gesture recognition for robot control. In J. S. Marques, N. Pérez de la Blanca, and P. Pina, editors, *Pattern Recognition and Image Analysis. IbPRIA 2005*, volume 3522 of *Lecture Notes in Computer Science*. Springer, Berlin, Heidelberg, 2005.

[111] N. Pavy, S. Rombauts, D. Déhais, C. Mathé, D. V. Ramana, P. Leroy, and P. Rouzé. Evaluation of gene prediction software using a genomic data set: Application to *Arabidopsis thaliana* sequences. *Bioinformatics*, 15(11):887–899, November 1999.

[112] S.-C. Poh, Y.-F. Tan, X. Guo, S.-N. Cheong, C.-P. Ooi, and W.-H. Tan. Lstm and hmm comparison for home activity anomaly detection. In *Proceedings of the IEEE 3rd Information Technology, Networking, Electronic and Automation Control Conference (ITNEC), Chengdu, China*, pages 1564–1568. IEEE, 2019.

[113] E. Prestat, M. M. David, J. Hultman, N. Taş, R. Lamendella, J. Dvornik, R. Mackelprang, D. D. Myrold, A. Jumpponen, S. G. Tringe, and E. Holman. Foam (functional ontology assignments for metagenomes): A hidden markov model (hmm) database with environmental focus. *Nucleic Acids Research*, 42(19):D145, 2014.

[114] R. Prytuliak, M. Volkmer, M. Meier, and B. H. Habermann. Hh-motif: De novo detection of short linear motifs in proteins by hidden markov model comparisons. *Nucleic Acids Research*, 45(1):470–477, 2017.

[115] L. R. Rabiner. A tutorial on hidden markov models and selected applications in speech recognition. *Proceedings of the IEEE*, 77:257–286, 1989.

[116] V. Radha and C. Vimala. A review on speech recognition challenges and approaches. *International Journal of Computer Science and Engineering*, 2(1):1–7, 2012.

[117] V. Ramasubramanian, A. K. V. Jayram, and T. V. Sreenivas. Language identification using parallel sub-word recognition—an ergodic hmm equivalence. In *Proceedings of Eurospeech, Geneva, Switzerland*, pages 1357–1360, 2003.

[118] D. Rapach, J. Strauss, and G. Zhou. Out-of-sample equity premium prediction: Combination forecasts and links to the real economy. *The Review of Financial Studies*, 23(2):821–862, 2010.

[119] D. A. Sadlier and N. E. O'Connor. Event detection in field sports video using audio visual features and a support vector machine. *IEEE Transactions on Circuits and Systems for Video Technology*, 15(10):1225–1233, 2005.

[120] R. San-Segundo, D. J. Echeverry-Correa, C. Salamea, and M. J. Pardo. Human activity monitoring based on hidden markov models using a smartphone. *IEEE Instrumentation & Measurement Magazine*, 19(6):27–31, 2016.

[121] B. Sanborn, D. Nguyen, and T. Nguyen. Global factor-based investing – opportunities. *NDR Explorer Insights Monthly*, Risks and Implementation, October 2017.

[122] D. Sanchez, M. Tentori, and J. Favela. Hidden markov models for activity recognition in ambient intelligence environments. In *Proceedings of the Eighth Mexican International Conference on Current Trends in Computer Science, Morelia, Mexico*, pages 33–40. IEEE, 2007.

[123] J. Sass and U. Haussmann. Optimizing the terminal wealth under partial information: The drift process as a continuous time markov chain. *Finance and Stochastics*, 8(4):553–577, 2004.

[124] G. Schwarz. Estimating the dimension of a model. *Annals of Statistics*, 6:461–464, 1978.

[125] F. Sha and L. K. Saul. Large margin hidden markov models for automatic speech recognition. In *Advances in Neural Information Processing Systems (NeurIPS)*, pages 1249–1256. MIT Press, 2007.

[126] A. Sloin and D. Burshtein. Support vector machine training for improved hidden markov modeling. *IEEE Transactions on Signal Processing*, 56(1):172–188, 2008.

[127] M. Taksar and X. Zeng. Optimal terminal wealth under partial information: Both the drift and the volatility driven by a discrete-time markov chain. *SIAM Journal on Control and Optimization*, 46(4):1461–1482, 2007.

[128] V. Ter-Hovhannisyan, A. Lomsadze, Y. Chernoff, and M. Borodovsky. Gene prediction in novel fungal genomes using an *Ab Initio* algorithm with unsupervised training. *Genome Research*, 18:1979–1990, 2008.

[129] S. M. Thede and M. P. Harper. A second-order hidden markov model for part-of-speech tagging. In *Proceedings of the 37th Annual Meeting of the Association for Computational Linguistics (ACL '99)*, pages 175–182, 1999.

[130] N. Thome and S. Miguet. A hhmm-based approach for robust fall detection. In *Proceedings of the 9th International Conference on Control, Automation, Robotics and Vision (ICARCV), Singapore*, pages 1–8. IEEE, 2006.

[131] N. Thome, S. Miguet, and S. Ambellouis. A real-time, multiview fall detection system: A lhmm-based approach. *IEEE Transactions on Circuits and Systems for Video Technology*, 18(11):1522–1532, 2008.

[132] M. K. Titsias, C. C. Holmes, and C. Yau. Statistical inference in hidden markov models using k-segment constraints. *Journal of the American Statistical Association*, 111(513):200–215, 2016.

[133] L. Tong, Q. Song, Y. Ge, and M. Liu. Hmm-based human fall detection and prediction method using tri-axial accelerometer. *IEEE Sensors Journal*, 13(5):1849–1856, 2013.

[134] K. Tra and T. V. Pham. Human fall detection based on adaptive background mixture model and hmm. In *Proceedings of the International Conference on Advanced Technologies for Communications (ATC 2013), Ho Chi Minh City, Vietnam*, pages 95–100. IEEE, 2013.

[135] D. Trabelsi, S. Mohammed, F. Chamroukhi, L. Oukhellou, and Y. Amirat. An unsupervised approach for automatic activity recognition based on hidden markov model regression. *IEEE Transactions on Automation Science and Engineering*, 10(3):829–835, 2013.

[136] D. Tran, W. Sheng, L. Liu, and M. Liu. A hidden markov model based driver intention prediction system. In *Proceedings of the IEEE International Conference on Cyber Technology in Automation, Control, and Intelligent Systems (CYBER), Shenyang, China*, pages 115–120. IEEE, 2015.

[137] S. Tu and L. Xu. A theoretical investigation of several model selection criteria for dimensionality reduction. *Pattern Recognition Letters*, 33:1117–1126, 2012.

[138] O.A. Vasicek. An equilibrium characterization of the term structure. *Journal of Financial Economics*, 5(2):177–188, 1977.

[139] N. Vaswani, A. K. Roy-Chowdhury, and R. Chellappa. Shape activity: A continuous-state hmm for moving/deforming shapes with application to abnormal activity detection. *IEEE Transactions on Image Processing*, 14(10):1603–1616, 2005.

[140] I. Visser and M. Speekenbrink. *Mixture and Hidden Markov Models with R*. Springer, 2022.

[141] A. J. Viterbi. Error bounds for convolutional codes and an asymptotically optimal decoding algorithm. *IEEE Transactions on Information Theory*, 13:260–269, 1967.

[142] C. Vogler and D. Metaxas. Adapting hidden markov models for asl recognition by using three-dimensional computer vision methods. In *Proceedings of the IEEE International Conference on Systems, Man, and Cybernetics. Computational Cybernetics and Simulation, Orlando, FL, USA*, volume 1, pages 156–161. IEEE, 1997.

[143] C. Wang, W. Gao, and S. Shan. An approach based on phonemes to large vocabulary chinese sign language recognition. In *Proceedings of the Fifth IEEE International Conference on Automatic Face and Gesture Recognition (AFGR), Washington, DC, USA*, pages 411–416. IEEE, 2002.

[144] J. Wang, C. Xu, E. Chng, and Q. Tian. Sports highlight detection from keyword sequences using hmm. In *Proceedings of the IEEE International Conference on Multimedia and Expo (ICME)*, volume 1, pages 599–602. IEEE, 2004.

[145] H. Wei, J. He, and J. Tan. Layered hidden markov models for real-time daily activity monitoring using body sensor networks. In *Proceedings of the 5th International Summer School and Symposium on Medical Devices and Biosensors (ISSMDBS), Hong Kong*, pages 326–329. IEEE, 2008.

[146] I. Welch and A. Goyal. A comprehensive look at the empirical performance of equity premium prediction. *The Review of Financial Studies*, 21(4):1455–1508, 2008.

[147] K. J. Won, T. Hamelryck, A. Prügel-Bennett, and A. Krogh. An evolutionary method for learning hmm structure: Prediction of protein secondary structure. *BMC Bioinformatics*, 8:1–3, December 2007.

[148] C. S. Wong and W. K. Li. On a mixture autoregressive model. *Journal of the Royal Statistical Society, Series B*, 62:95–115, 2000.

[149] P. C. Woodland and D. Povey. Large scale discriminative training of hidden markov models for speech recognition. *Computer Speech & Language*, 16(1):25–47, 2002.

[150] J. Wu and J. Xie. Hidden markov model and its applications in motif findings. *Statistical Methods in Molecular Biology*, pages 405–416, 2010.

[151] L. Xie, S. F. Chang, A. Divakaran, and H. Sun. Unsupervised discovery of multilevel statistical video structures using hierarchical hidden markov models. In *Proceedings of the IEEE International Conference on Multimedia and Expo (ICME)*, pages III–29. IEEE, 2003.

[152] L. Xie, P. Xu, S. F. Chang, A. Divakaran, and H. Sun. Structure analysis of soccer video with domain knowledge and hidden markov models. *Pattern Recognition Letters*, 25:767–775, 2004.

[153] C. Xiong, D. Yang, and L. Zhang. A high-order hidden markov model and its applications for dynamic car ownership analysis. *Transportation Science*, 52(6):1365–1375, 2018.

[154] G. Xu, Y. F. Ma, and S. Yang. A hmm based semantic analysis framework for sports game event detection. In *Proceedings of the IEEE International Conference on Image Processing (ICIP)*, volume 1, pages I–25. IEEE, 2003.

[155] M. Xu, L. Zuo, S. Iyengar, A. Goldfain, and J. DelloStritto. A semi-supervised hidden markov model-based activity monitoring system. In *Proceedings of the Annual International Conference of the IEEE Engineering in Medicine and Biology Society (EMBC), Boston, MA, USA*, pages 1794–1797. IEEE, 2011.

[156] J. Yamato, J. Ohya, and K. Ishii. Recognizing human action in time-sequential images using hidden markov model. In *Proceedings of the IEEE Computer Society Conference on Computer Vision and Pattern Recognition (CVPR)*, pages 379–385. IEEE, 1992.

[157] S.-H. Yang, K. Thapa, M. H. Kabir, and H.-C. Lee. Log-viterbi algorithm applied on second-order hidden markov model for human activity recognition. *International Journal of Distributed Sensor Networks*, 14(4):1550147718772541, 2018.

[158] M. Yeluri. Understanding and implementing speech recognition using hmm. `https://maharshi-yeluri.medium.com/understanding-and-implementing-speech-recognition-using-hmm-6a4e7666de1`. Accessed: 2025-11-03.

[159] D. Yu and L. Deng. Large-margin discriminative training of hidden markov models for speech recognition. In *Proceedings of the International Conference on Semantic Computing*, pages 429–438, 2007.

[160] S. Yu, H. Chen, and R. A. Brown. Hidden markov model-based fall detection with motion sensor orientation calibration: A case for real-life home monitoring. *IEEE Journal of Biomedical and Health Informatics*, 22(6):1847–1853, 2018.

[161] Y. Yuan, X. Li, Z. Liu, and X. Guan. Occupancy estimation in buildings based on infrared array sensors detection. *IEEE Sensors Journal*, 20(2):1043–1053, January 2020.

[162] S. A. Zenios. A model for portfolio management with mortgage-backed securities. *Annals of Operations Research*, 43(6):337–356, 1993.

[163] X. Zhu and J. Zhu. Predicting stock returns: A regime-switching combination approach and economic links. *Journal of Banking & Finance*, 37(11):4120–4133, 2013.

[164] Y. A. Zhu and J. Cheng. Using hidden markov model to detect macro-economic risk level. *Review of Integrative Business & Economics Research*, 2(1):238–249, 2013.

[165] W. Zucchini and I. L. MacDonald. *Hidden Markov Models for Time Series: An Introduction Using R.* Chapman and Hall/CRC, 2009.

Index